Crystal Technology

WILEY SERIES IN PURE AND APPLIED OPTICS

Advisory Editor

Stanley S. Ballard, University of Florida

ALLEN AND EBERLY • *Optical Resonance and Two-Level Atoms*

BOND • *Crystal Technology*

CATHEY • *Optical Information Processing and Holography*

CAULFIELD AND LU • *The Applications of Holography*

FRANCON AND MALLICK • *Polarization Interferometers*

GERRARD AND BURCH • *Introduction to Matrix Methods in Optics*

HUDSON • *Infrared System Engineering*

JUDD AND WYSZECKI • *Color in Business, Science, and Industry,* Third Edition

KNITTL • *Optics of Thin Films*

LENGYEL • *Introduction to Laser Physics*

LENGYEL • *Lasers,* Second Edition

LEVI • *Applied Optics, A Guide to Optical System Design,* Volume I

LOUISELL • *Quantum Statistical Properties of Radiation*

MOLLER AND ROTHSCHILD • *Far-Infrared Spectroscopy*

PRATT • *Laser Communication Systems*

SHULMAN • *Optical Data Processing*

WILLIAMS AND BECKLUND • *Optics*

ZERNIKE AND MIDWINTER • *Applied Nonlinear Optics*

Crystal Technology

W. L. BOND
Stanford University

John Wiley & Sons New York/London/Sydney/Toronto

CHEMISTRY

Library of Congress Cataloging in Publication Data:

Bond, Walter Lysander, 1903–
    Crystal technology.

    (Wiley series in pure and applied optics)
    Includes index.
    1. Crystallography. I. Title.
QD905.2.B68 1976   548   75-23364
ISBN 0-471-08765-3

Printed in the United States of America

10 9 8 7 6 5 4 3 2 1

Preface

Crystal materials, so important to modern electronics and optics, are often grown and processed by people who have only minimal knowledge about crystals and methods of cutting them to various shapes and with specified orientation. This field lies at the point where optics, geometry, crystallography, and instrument making intersect, and this book is written to throw some light on this intersection.

The work is organized as follows. After a study of crystal symmetry, crystallographic axes, and unit cells, we show how atomic planes are specified. We then treat the specifying of crystal plate orientations in terms of crystallographic axes and means of realizing such orientations with the help of x-rays, light, polarized light, etch pits, and piezoelectric polarity testers. This is followed by a treatment of sawing, flat grinding, parallel grinding, prism making, spherical grinding, cylindrical grinding, and the making of surfaces of double curvature (toroids mainly). Instruments are described for testing these and for testing crystal perfection. Also included is a list of x-ray Bragg angles for some 70 crystals of interest in present-day solid state physics.

A concluding chapter gives a peek into methods for studying light refraction in crystals in general and for calculating sound velocity in any direction in any crystal for which the elastic moduli are known.

I thank Miss Nancy Ennis and Ms. Evelyn Morris for typing the manuscript, and especially Ms. Audrey Kenney for making the many fine drawings.

W. L. BOND

Stanford, California
July 1975

v

Contents

Crystal Technology

Crystal Technology
by W. L. Bond

ERRATA

Page 8. Figure 1.19 should appear as follows:

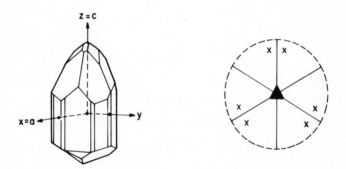

Figure 1.19 A tourmaline crystal and its symmetry diagram, hexagonal (trigonal), class C_{3v}-$3m$.

1

Crystal Systems and Orientation of Plates

CRYSTAL SPECIFICATION: CLASSIFICATION

In well-annealed glass all directions are equivalent in the sense that the gross physical properties in one direction are the same as those in any other direction. By gross physical properties we mean such things as sound velocity and velocity of light; we exclude atomic scale phenomena, however. The atoms in glass have only slight local order, but this degree of order is statistically the same in all parts of the piece.

Crystals have a high degree of order, not only locally but throughout; that is, crystals are orderly arrangements of atoms but the order may not be simple. Yet in any single crystal we can isolate a unit parallelopiped, from which the entire crystal can be imagined to be constructed, by close stacking of such units in three dimensions. These units, called unit cells, are all oriented the same way; each looks the same as the units are assembled in the crystal. Because of this order there may be several equivalent directions in a crystal, directions for which the sound velocities are equal, the light velocities are equal, and all other physical properties are the same in these several equivalent directions.

X-ray crystallographers have analyzed many thousands of crystals and have determined the size and shape of the unit cell for each crystal (i.e., the lattice constants) as well as the arrangement of atoms in each unit cell.

Most natural crystals look the same from several directions; that is, they have symmetry. Mineralogists had noted this before X-rays came into use and had divided crystals into seven crystal systems according to this gross symmetry. The systems were called: triclinic, monoclinic, orthorhombic, tetragonal, trigonal, hexagonal, and cubic. The symmetry elements they used were axes of twofold, threefold, fourfold, and sixfold symmetry. Also they observed planes of mirror imagery and centers of

1

inversion. Threefold symmetry is now often considered to be a special case of sixfold symmetry. It can be shown that no other kinds of axes can be used if the crystal is still to be considered as made by the close stacking of similarly oriented unit cells. Axes of fivefold or eightfold symmetry, and so on, do not exist in crystals.

A crystal can have several kinds of such permitted axes simultaneously. We can have three mutually perpendicular axes of fourfold symmetry and, at the same time, four threefold axes all intersecting at a point; also several mirror planes may pass through this point. Many such crystals exist. By making all the logically possible combinations of these symmetry elements, we arrive at 32 possible classes called "point groups."

The 32 Point Groups

The least symmetry a crystal can have is that of the triclinic asymmetric class, which has no symmetry. The atoms in the unit cell are arranged without symmetry. The cell is, in general, a nonrectangular parallelopiped. To describe the cell we choose vectors a, b, c from an origin at a cell corner. The lengths of these are a_0, b_0, and c_0, respectively, and these lengths are unequal. The angle between vectors b and c is called α, that between c and a is β, that between a and b is γ.

In general a, b, c are chosen such that c_0 is the shortest, b_0 is the longest; also α and β must be obtuse.

It is possible to arrange atoms within a triclinic cell so that there is a center of symmetry. This gives us the class "triclinic pinacoidal." There are no other classes in the triclinic system.

Before proceeding with the other crystal systems we introduce two competing systems of names for the point groups, the Schoenflies notation, and the Hermann-Mauguin notation. Using Schoenflies, for the triclinic asymmetric class, we write C_1-1. The 1 as a subscript of the C means that there is only a onefold symmetry axis; that is, only a complete turn about an axis makes the crystal look as it did before turning. The 1 following the dash means the same thing in Hermann-Mauguin nomenclature. For the triclinic pinacoidal point group we write C_i-$\bar{1}$. The subscript i denotes inversion as the only symmetry element, the $\bar{1}$ means a complete turn about an axis has occurred, plus an inversion.

A word about inversion: if every property in a crystal is the same in either direction along any line in the crystal, the crystal is said to possess a center of inversion. Each face of the axinite crystal in Fig. 1.2 is matched by a similar face on the opposite side of the crystal. Such a crystal could not exhibit a piezoelectric effect. Distortion of the crystal can only cause equal electric charges to move equally in opposite directions along any line in the crystal. Hence no electric field can result.

In the monoclinic system the b axis is perpendicular to the a and c axes; that is, $\alpha = \gamma = 90°$ but $\beta > 90°$. The lengths a_0, b_0, and c_0 are still unequal. Now either b is a binary axis (i.e., one of twofold symmetry), or it is perpendicular to a plane of reflection symmetry (commonly called a mirror plane and denoted by m), or it is both binary and perpendicular to a mirror plane (i.e., three classes). It is evident that the symbol for a monoclinic class with only a binary axis is C_2-2, and the monoclinic class with a mirror plane only is denoted by C_s-m. The class having both a binary axis and a perpendicular mirror plane is designated by C_{2h}-$2/m$. The slash line in the Hermann-Mauguin part indicates that the mirror plane is perpendicular to the binary axis.

In the orthorhombic system $\alpha = \beta = \gamma = 90°$, but $a_0 \neq b_0 \neq c_0$. Here we may have three mutually perpendicular binary axes, class D_2-222, or we can have two mirror planes intersecting at right angles, the line of intersection being also a binary axis—class C_{2v}-$2mm$; finally, an orthorhombic crystal may have three mutually perpendicular mirror planes with the three lines of intersection being binary axes, class D_{2h}-mmm.

The tetragonal system introduces a new factor. The simplest tetragonal symmetry is an axis whereby an equivalent direction is given by a quarter turn plus an inversion. This is class S_4-$\bar{4}$. A tetragonal crystal with only a simple fourfold axis is of class C_4-4.

Instead of word descriptions of the other 22 classes we present maps of the symmetry elements of all crystal classes (Figs. 1.1 through 1.32). The small crosses and circles show equivalent directions. A cross can be thought of as D degrees above the plane of projection and the circle as D degrees below the plane of projection. We also illustrate a natural crystal in each class. Here a boat-shaped figure indicates a binary axis, a solid triangle indicates a trigonal (threefold) axis, and a solid square indicates a

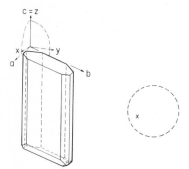

Figure 1.1 An aminoethyl ethanolamine hydrogen tartrate crystal and its symmetry diagram, triclinic, class C_1-1.

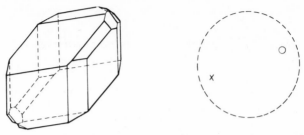

Figure 1.2 An axinite crystal and its symmetry diagram, triclinic, class C_i-$\bar{1}$.

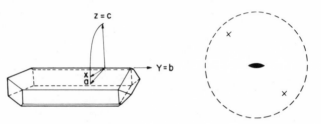

Figure 1.3 An ethylene diamine tartrate crystal and its symmetry diagram, monoclinic, class C_2-2.

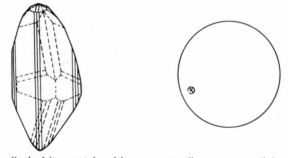

Figure 1.4 A clinohedrite crystal and its symmetry diagram, monoclinic, class C_s-m.

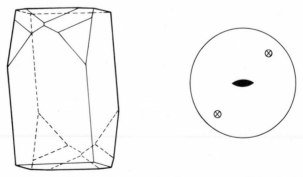

Figure 1.5 A pyroxene crystal and its symmetry diagram, monoclinic, class C_{2h}-$2/m$.

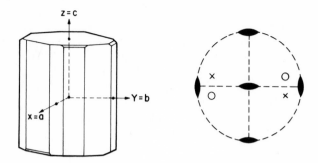

Figure 1.6 A rochelle salt crystal and its symmetry diagram, orthorhombic, class D_2-222.

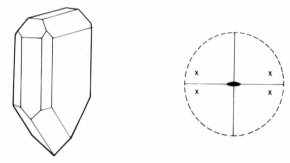

Figure 1.7 A calamine crystal and its symmetry diagram, orthorhombic, class C_{2v}-$2mm$.

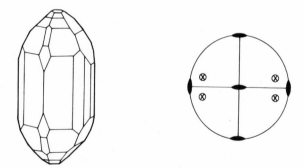

Figure 1.8 A topaz crystal and its symmetry diagram, orthorhombic, class D_{2h}-mmm.

fourfold axis; but a hollow square containing a boat-shaped figure indicates a fourfold inversion axis $\bar{4}$. A hollow hexagon containing a solid triangle indicates a sixfold inversion axis $\bar{6}$, whereas a solid hexagon indicates a sixfold axis. A solid circle shows the equatorial plane to be a mirror plane, and the other solid lines indicate other mirror planes.

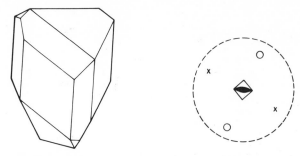

Figure 1.9 An idealized cahnite crystal and its symmetry diagram tetragonal, class S_4-$\bar{4}$.

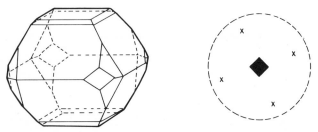

Figure 1.10 A wulfenite crystal and its symmetry diagram, tetragonal, class C_4-4.

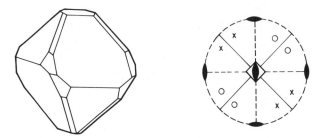

Figure 1.11 A chalcopyrite crystal and its symmetry diagram, tetragonal, class D_{2d}-$\bar{4}2m$.

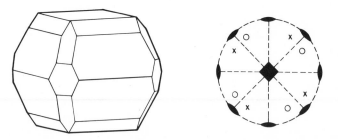

Figure 1.12 A crystal of nickel sulfate and its symmetry diagram, tetragonal, class D_4-422.

6

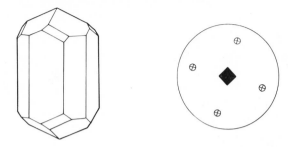

Figure 1.13 A crystal of wernerite and its symmetry diagram, tetragonal, class C_{4h}-$4/m$.

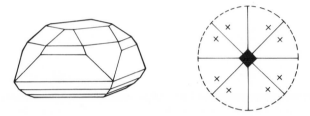

Figure 1.14 A crystal of diabolite and its symmetry diagram, tetragonal, class C_{4v}-$4/mm$.

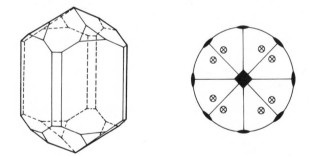

Figure 1.15 A crystal of rutile and its symmetry diagram, tetragonal, class D_{4h}-$4/mmm$.

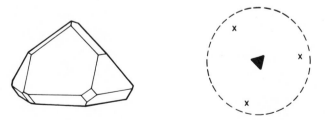

Figure 1.16 A crystal of sodium periodate and its symmetry diagram, hexagonal (trigonal), class C_3-3.

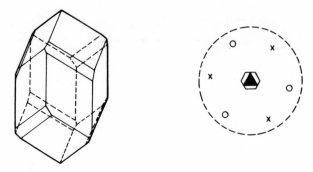

Figure 1.17 A crystal of dioptase and its symmetry diagram, hexagonal (rhombohedral), class C_{3i}-$\bar{3}$.

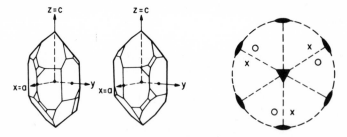

Figure 1.18 Two crystals of quartz: left-handed and right-handed with symmetry diagram, hexagonal (trigonal), class D_3-32.

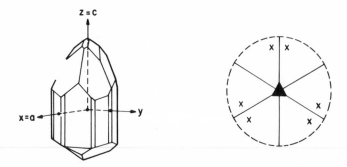

Figure 1.19 A tourmaline crystal and its symmetry diagram, hexagonal (trigonal), class C_{3v}-3m.

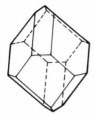 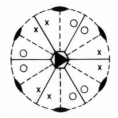

Figure 1.20 A crystal of calcite and its symmetry diagram, hexagonal (rhombohedral), class $D_{3d}\text{-}\bar{3}m$.

NO CERTAIN EXAMPLE

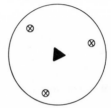

Figure 1.21 Symmetry diagram of class $C_{3h}\text{-}\bar{6}$, hexagonal (no certain example known).

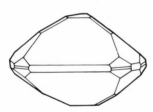 

Figure 1.22 A benitoite crystal and its symmetry diagram, hexagonal (trigonal), class $D_{3h}\text{-}\bar{6}m\,2$.

Figure 1.23 A nepheline crystal and its symmetry diagram, hexagonal, class $C_6\text{-}6$.

No certain example with natural
faces has been found

Figure 1.24 Hexagonal, class D_6-622 symmetry diagram (no certain example with natural faces observed).

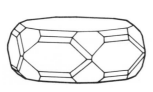 

Figure 1.25 An apatite crystal and its symmetry diagram, hexagonal, class C_{6h}-6/m.

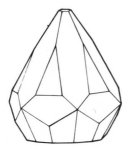 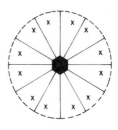

Figure 1.26 A zincite crystal and its symmetry diagram, hexagonal, class C_{6v}-6mm.

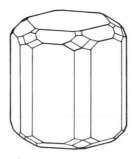 

Figure 1.27 A beryl crystal and its symmetry diagram, hexagonal, class D_{6h}-6/mmm.

10

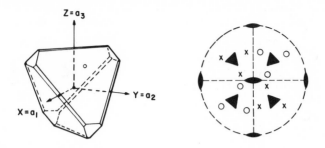

Figure 1.28 A right-handed crystal of sodium bromate and its symmetry diagram, cubic, class T-23.

NO CERTAIN EXAMPLE

Figure 1.29 Symmetry diagram of cubic, class O-432 (no certain example is known).

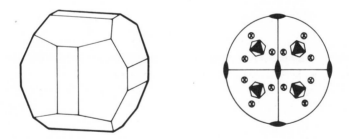

Figure 1.30 A pyrite crystal and its symmetry diagram, cubic, class T_h-$m3$.

An understanding of the symmetry of crystals guides us in alternative ways for cutting equivalent specimens from a raw crystal. This knowledge is important in cutting synthetic crystals; of two equivalent directions one may lie across a long boule, the other along the length of the boule. For a long plate we make the length lie along the length of the boule.

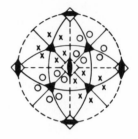

Figure 1.31   A sphalerite crystal and its symmetry diagram, cubic, class T_d-$\bar{4}3m$.

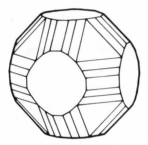
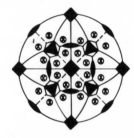

Figure 1.32   A galena crystal and its symmetry diagram, cubic, class O_h-$m3m$.

PLATE SPECIFICATIONS

For mechanical convenience we now define unitary orthogonal axes for all crystals. We take Z along the c axis and X in the plane of the c and a axes, placing Y to form a right-handed coordinate system. Figure 1.1 illustrates this for a triclinic crystal. The Y axis is always perpendicular to the face marked 010.

Unrotated or Simple Orientations

We specify simple cuts by a simple orientation symbol such as (xy) where x, y, z refer to the above-mentioned X, Y, Z axes, respectively. For very high frequency oscillations the smallest dimension is the most important, hence the thickness direction is mentioned first in the specification. Since length is the most important dimension for low frequencies, the length direction is mentioned next in the plate specification. That is, an (xy) cut has its thickness direction along the X axis and its length along the Y axis, as in Fig. 1.33. Continuing in this way, Figs. 1.34 through 1.38 illustrate (xz), (yx), (yz), (zx), and (zy) orientations.

Figure 1.33   An (xy) cut.

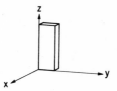

Figure 1.34   An (xz) cut.

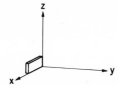

Figure 1.35   A (yx) cut.

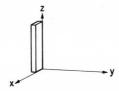

Figure 1.36   A (yz) cut.

Figure 1.37   A (zx) cut.

Figure 1.38   A (zy) cut.

Singly Rotated Orientations

Plates oriented in the simple way just described often do not give the best possible performance for a particular use. Better results follow if the plate is cut out of the mother crystal at an angle to one of the six simple orientations. For example, the AT quartz plate has a zero temperature coefficient of frequency because it is rotated $35°15'$ from the (yx) or (yz) simple orientation. We acknowledge that this rotation is about the length direction from a starting (yx) orientation by calling this a (yxl) ϕ orientation, where ϕ is the counterclockwise rotation of the plate, indicated in Fig. 1.39. Here ϕ may be positive or negative. Figure 1.40 shows the analogous rotation (yzt) ϕ. There are 16 other possible simple rotation symbols, since each of the initial positions xy, xz, yx, yz, zx, zy can be rotated about the length direction, the width direction, or the thickness direction (i.e., 18 possibilities all together). These should be obvious and need not be illustrated.

Figure 1.39 A (yzw) 30° (positive) rotation.   **Figure 1.40** A (yzt) ϕ rotation.

Many crystals are cut into square plates or round disks. When this is done the length or width is indistinguishable and the symbol used is somewhat arbitrary. For example, an AT cut quartz plate could be either a (yzw) 35°15′ or a (yxl) 35°15′ orientation.

Doubly Rotated Orientations

After a simple rotation, a second rotation is sometimes necessary to achieve improved performance. The notation is accordingly extended by adding a second letter, l, t, or w to the first l, t, or w inside the parentheses and adding a second angle at the end. Orientation $(yztw)$ 30°, 40° appears in Fig. 1.41. Of course the second l, t, or w will be different from the first one because, for example, an orientation $(yztt)$ 10°, 20° would be more simply a (yzt) 30° orientation.

Triply Rotated Orientations

All practical problems can be solved with no more than three rotations. The nomenclature is therefore extended by adding a third l, w, or t inside the parentheses and adding a third angle outside at the end. The third l, w, or t can be the same as the first but not the same as the second. In Fig. 1.42 we have a $(yztwt)$ 30°, 40°, 15° orientation given by further rotation of the example of Fig. 1.41 a, b, c by 15° about t. The sense of all these rotations is counterclockwise, as seen looking toward the origin from the positive end of the x, y, z axes (i.e., positive if the rotation angle is stated as positive).

Three rotations such as those just described can produce any orientation from any starting position. In fact any orientation can be produced from any starting position with just one rotation if we are allowed to specify an axis not necessarily along a coordinate axis direction. However the method described is easier conceptually and easier to execute with available mechanical tools.

In making rotated cuts such as that illustrated in Fig. 1.39, we need to know the positive end from the negative end of the coordinate axes x, y, z. In Fig. 1.39 if $-x$ is mistaken for $+x$, the angle 30° becomes effectively $-30°$ and the properties of a $-30°$ plate may be quite different from those

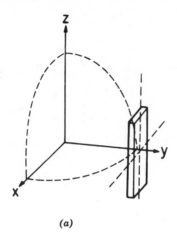

(a)

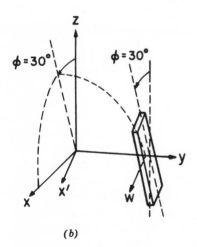

(b)

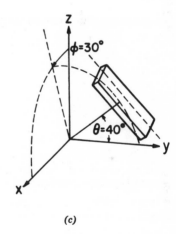

(c)

Figure 1.41 A (*yztw*) 30°, 40° rotation.

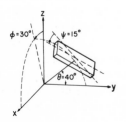

Figure 1.42 A (*yztwt*) 30°, 40°, 15° rotation.

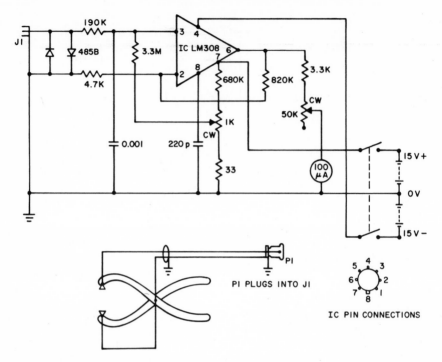

Figure 1.43 W. O. Heintzen's piezoelectric polarity tester.

of a $+30°$ plate. If the plate is cut from a crystal with many natural faces, these may indicate the plus from the minus axes. For example, Fig. 1.18 shows left quartz in the left-hand scheme, right quartz on the right. The positive ends of the two x axes are as illustrated. The x axis is an axis of binary symmetry for both left and right. Hence either end of y and z may be considered as the plus end, but the three axes must form a right-hand system.

If the raw material does not have natural faces, one needs help from X-rays, polarized light, and possibly a piezoelectric squeeze tester. Polarized light will show whether the piece is left or right quartz. (In the conoscope, Chapter 5, the interference rings expand for right quartz if the analyzer is turned clockwise; they contract for left quartz.) For right quartz the piezoelectric constant d_{11} is negative. For left quartz it is positive. On compression along the x axis, the positive end of the x axis of right quartz becomes electrically positive, for left quartz, negative. This can be determined by means of a vacuum tube voltmeter or the integrated circuit test meter of W.O. Heintzen (Fig. 1.43).

2

Atomic Planes and their Indices

ATOMIC PLANES

A natural crystal is often bounded by plane faces as we saw in Figs. 1.1 through 1.32. Some of these faces are considered to be flat layers of unit cells (left-hand part of Fig. 2.1). If this were the only kind of surface, crystals bounded by flat surfaces could be only parallelopipeds. Stepped surfaces (right-hand part of Fig. 2.1) must exist as well as doubly stepped surfaces, shown in Fig. 2.2. In fact the rule "over one cell then back one cell and repeat" may be replaced by "over m cells then back n cells," where m and n are small integers, generally not greater than 4. For surfaces more general than that in Fig. 2.2 we need three small integers, say m, n, and p, each commonly not larger than 4, to describe all natural faces. Measurements of interfacial angles confirm that all such angles can be explained in terms of a unit cell of proper shape and small integers combined appropriately.

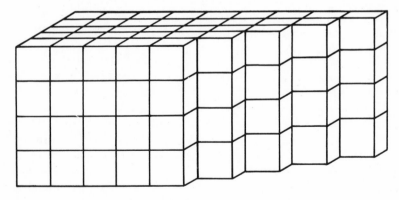

Figure 2.1 Two kinds of atomic plane.

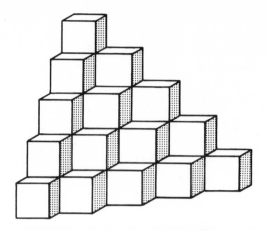

Figure 2.2   A third kind of atomic plane.

We said originally that it was always possible to choose a unit cell from which the crystal could be imagined to be made by "same-orientation close stacking." There are many ways in which a unit cell can be chosen, however; hence there are many ways of defining the crystal axes. The specification of atomic planes depends on the choice of axes. It is customary to choose the smallest unit cell that has the symmetry of the crystal, but there are exceptions to this rule.

MILLER INDICES

Instead of describing planes of atoms in terms of steps m, n, and p, it has been found useful to define atomic planes in terms of "Miller indices" (h, k, l), where h, k, and l are small integers. The plane (hkl) has intercepts on the crystal axes a, b, c of a/h, b/k, and c/l, respectively; that is, the plane cuts the a axis at $(1/h)$th of its length, the b axis at $(1/k)$th of its length, and the c axis at $(1/l)$th of its length. This is illustrated in Fig. 2.3.

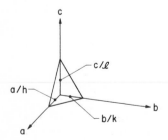

Figure 2.3   Miller indices of plane (hkl).

It will be seen that a plane (100) is parallel to the b and c axes, a plane (010) is parallel to the c and a axes, and a plane (001) is parallel to the a and b axes. Also a plane ($0kl$) is parallel to the a axis, a plane ($h0l$) is parallel to b, and a plane ($hk0$) is parallel to c. This is illustrated in epidote (Fig. 2.4), a monoclinic crystal. Here face $c = (001)$ is parallel to the b

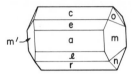

Figure 2.4   Epidote.

axis, as also is face $l = (\bar{2}01)$. The bar over the 2 designates -2, to avoid writing an expression as awkward as $(1-21)$; clearly $(1\bar{2}1)$ is more compact. Face $a = (100)$ is parallel to the b and c axes. Continuing, we observe that $r = (\bar{1}01)$ is parallel to the b axis, face c is parallel to the a and b axes as stated previously, face $m = (110)$ is parallel to the c axis, as is also face $m' = (1\bar{1}0)$.

MILLER–BRAVAIS INDICES

In the hexagonal system there are several equally advantageous ways of choosing three axes, as indicated by Fig. 2.5. We could take a_1, a_2, c, or a_2, a_3, c, and so on, since a_1, a_2, a_3 are equivalent directions. We could not choose a_1, a_2, a_3 without c because a_1, a_2, a_3 are coplanar and could not be edges of a three-dimensional unit cell. There is an advantage in using all four axes. Bravais introduced four indices $hkil$. The meaning of these indices is given in Fig. 2.6. The advantage is that the sum of hk and i is always zero, hence $i = -(h + k)$ and the plane ($ihkl$) is equivalent to the plane ($hkil$), as is the plane $kihl$. That is, h, k, and i may be cycled to give equivalent planes. For example, there is often a face $(51\bar{6}1)$ on natural

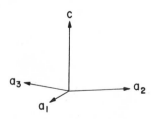

Figure 2.5   Hexagonal axes.

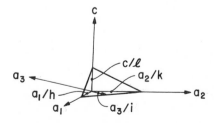

Figure 2.6   Bravais indices of a plane.

quartz crystals and there is also a face ($\bar{6}511$) and a face ($1\bar{6}51$). In general ($hikl$) is not a plane equivalent to ($hkil$); h, k, and i must be kept in their cyclic order.

Since i is always equal to $-(h + k)$, it is easily supplied and may be omitted. However, it is safer to indicate this omission by a dot as ($hk \cdot l$) unless it is specifically stated that (hkl) is on hexagonal coordinates. We take this precaution because there is a rhombohedral system that can be converted to hexagonal coordinates, and many crystallographers have stated their results on rhombohedral axes.

RHOMBOHEDRAL AXES

Hexagonal axes have many advantages over rhombohedral axes. There are effectively two ways of relating rhombohedral indices to hexagonal indices, the obverse and the reverse relationship. The rhombohedral unit cell in Fig. 2.7 has edges r_1, r_2, and r_3, each at an angle α with the other two axes. We see that the rhombohedral cell is like a cube that has been stretched or compressed along one body diagonal. In fact a cube can be considered to be a very special case of a rhombohedral crystal.

For obverse hexagonal axes we take hexagonal a_1 parallel to a line joining the terminus of r_3 to the terminus of r_2. Hexagonal a_2 is taken parallel to the line joining the terminus of r_1 to the terminus of r_3, and a_3 is taken parallel to the line joining the terminus of r_2 to the terminus of r_1.

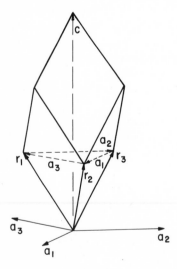

Figure 2.7 Obverse relation of rhombohedral axes converted to hexagonal axes.

The c axis is the body diagonal of the rhombohedral cell. The hexagonal unit cell is defined by a_1, a_2, and c. It has a larger volume than the rhombohedral cell, hence contains more atoms. The "lattice constants" of the rhombohedral cell are a cell edge and α. We have called a cell edge r for the rhombohedral cell, but a for the related hexagonal cell.

Given r_0 the length of a rhombohedral axis and α for a rhombohedral cell, we find the length of the hexagonal axis a_0 by means of the equation

$$a_0 = 2r_0 \sin(\alpha/2) \qquad (2.1)$$

as shown by inspection of Fig. 2.7. The hexagonal c axis is given by

$$c_0^2 = 9r_0^2 - 3a_0^2 \qquad (2.2)$$

Inversely we write

$$r_0^2 = \frac{a_0^2}{3} + \frac{c_0^2}{9} \qquad (2.1')$$

$$\sin^2(\alpha/2) = \frac{9}{4(3 + c_0^2/a_0^2)} \qquad (2.2')$$

The change of axes causes a change of the plane intercepts, hence requires a change in plane indices. A change of axes from a, b, c, to a', b', c', where $a' = u_1a + v_1b + \omega_1c$, $b' = u_2a + v_2b + \omega_2c$, and $c' = u_3a + v_3b + \omega_3c$ changes Miller indices hkl to $h'k'l'$, where $h' = u_1h + v_1k + l_1c$, $k' = u_2h + v_2k + \omega_2l$, and $l' = u_3h + v_3k + \omega_3l$.

We call indices on the rhombohedral axes (mnp) and on the hexagonal axes $(hk\cdot l)$. Hence we have

$$\begin{aligned} h &= & n - p \\ k &= -m & + p \\ l &= & m + n + p \end{aligned} \qquad (2.3)$$

Note that for this relation between rhombohedral axes and hexagonal axes (i.e., the obverse relation) we have

$$-h + k + l = 3p \qquad (2.4)$$

a number divisible by 3.

There is a second relation called the reverse relation in which a_1 and a_2 are reversed in direction. For this we find

$$h - k + l = 3p \qquad (2.4')$$

The restrictions of Eqs. 2.4 and 2.4' tell us that certain hkl values are meaningless in the sense that these planes do not correspond to planes of atoms. Using the first convention we find that for rhombohedral crystals, planes $10\cdot0$, $12\cdot0$, $13\cdot0$ are meaningless as are such planes as $10\cdot2$, $10\cdot3$, and

10·5, whereas 01·5, 01·8, 11·0, 11·3, 12·2, 10·7, and so on, are valid. With the reverse relationship, plane 01·5 is not valid but 10·5 is valid, 12·2 is not valid but 21·2 is valid—an interchange of the first two figures takes us from a plane that is valid in the obverse unit cell to one valid in the reverse unit cell.

The relation between rhombohedral cells and hexagonal cells is paralleled by several other such relationships. To have the unit cell edges parallel to axes of symmetry or the cell faces parallel to mirror planes, we may place an atom group at the cell center, at the center of all faces, or else at only two opposite faces. If there are atom groups only at cell corners, the cell is called primitive and is symbolized by the letter P. If there is also one centered on each face it is called face centered, designated by F. If there is such a group at each corner and one in the center of the cell, it is called body centered or inner centered, designated by I. If only one pair of opposite faces has an atom group at its center, the cell is said to be base centered, designated by an A, B, or C, according to which faces are centered. We find that we cannot center two sets of faces and build a crystal by translations only.

The centerings just named can be removed by choosing new unit cells if we ignore the symmetry of the crystal. For a primitive cell all integral values of h, k, and l are valid. We can remove the body centering if we choose a new cell in which there is no centering. When we do this and compute h, k, l on the old body-centered cell, we find that $h + k + l$ must be an even number. This is generally stated as $h + k + l = 2n$. Similarly a face-centered cell can be made primitive by picking a new set of axes going from the origin to the three nearest face centers. This gives a primitive cell in which all h's, k's, and l's are valid; but when we return to the old face-centered cell we find that the new h's, k's, l's are all odd or all even. If the cell is base centered on the faces, parallel to the a and b axes (i.e., C centered), it can be shown that $h + k$ must be even. Centering on the faces parallel to b and c (i.e., A centering) requires that $k + l$ be even, and centering on the faces parallel to the axes c and a (i.e., B centering) calls for l and h to be even.

PROOF THAT A TRANSLATIONAL LATTICE CAN HAVE ROTATIONAL SYMMETRY AXES OF MULTIPLICITY 1, 2, 3, 4, AND 6 ONLY

Given a translational lattice L, if there is an n-fold axis of rotational symmetry at a point a_1 (Fig. 2.8), translation produces others. We assume that at point b_1 (Fig. 2.8) is another rotational symmetry axis just like the one at point a_1 and that there are no others like these closer together than the pair a_1, b_1.

Since a_1 is an n-fold axis, all points such as point b_1 must be repeated

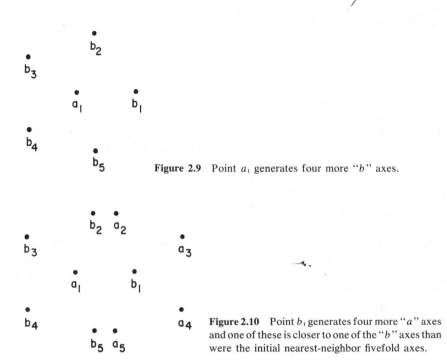

Figure 2.8 Two equivalent hypothetical "nearest neighbor" fivefold axes.

about a_1 to give n points like a_1 and b_1 all equidistant from point a_1. This is illustrated in Fig. 2.9 for $n = 5$. However point b_1 is also an n-fold symmetry axis; thus there must be n points like point a_1 all equidistant from point b_1, as in Fig. 2.10. Now the points a_2 and b_2 are much closer

Figure 2.9 Point a_1 generates four more "b" axes.

Figure 2.10 Point b_1 generates four more "a" axes and one of these is closer to one of the "b" axes than were the initial nearest-neighbor fivefold axes.

together than were points a_1 and b_1. But this violates the assumption that there were no such axes closer together than points a_1 and b_1. We could now start with points a_2 and b_2 and repeat the foregoing construction to show that for $n = 5$ there must be points of fivefold symmetry even closer together than points a_2 and b_2. Continuing in this way we can show that there can be no closest distance between these axes; that is, we have a continuum, not a discrete lattice.

If $n = 2, 3, 4,$ or 6, points such as a_1 and b_1 would coincide and there would be no contradiction. For $n = 8$ the contradiction occurs, as it does for all numbers except 1, 2, 3, 4, and 6.

3

Space Groups and X-Ray Reflection

Before X-rays were used, the shape of the cell could be deduced; that is, its angles α, β, γ and the ratio of the axial lengths could be evaluated. Since the absolute lengths of a, b, and c were unknown, the ratio was stated as $a : b : c$, with b set at 1.0000. For example, titanite was given as monoclinic $a : b : c = 0.7547 : 1 : 0.8543$, $\beta = 60°17'$. Nowadays we turn the crystal around to make β obtuse and find X-ray data as $a_0 = 6.55$ Å, $b_0 = 8.70$ Å, $c_0 = 7.43$ Å, $\beta = 119°43'$, where Å is the symbol for angstrom unit, 10^{-8} cm. This gives axial ratios $0.7529 : 1 : 8540$. All the old data are from measurement of interfacial angles. The discrepancy between 0.7547 and 0.7529 could be due to impurities, which change the cell dimensions. However many of the data cannot be so simply converted because X-rays often reveal a cell smaller than that assumed by the pre-X-ray mineralogist. As an example of this we consider the case of corundum, α aluminum oxide. The old mineralogy gave the data merely as the c_0/a_0 ratio, called $c = 1.3630$. The X-ray data are given as $r_0 = 5.12$ Å, $\alpha = 55°17'$. From this, Eqs. 2.1 and 2.2 give $a_0 = 4.7507$, $c_0 = 12.970$. This yields an axial ratio $c_0/a_0 = 2.730$, about twice the old mineralogical result. How then is the new cell smaller than the old cell? Equation 2.2′ gives $\sin(\alpha'/2) = 3/2\sqrt{3 + 1.363^2}$, whence $\alpha' = 85°47'$, a fatter cell but less tall than the X-ray cell, for which $\alpha = 57°17'$.

Space Groups

The 32 point groups or crystal classes were derived using axes of symmetry, axes of inversion symmetry, and planes of reflection symmetry. These elements were sufficient to explain all interfacial angles of any crystal with the assumption of a unit cell of a certain shape for any

24

given crystal. Yet these symmetry elements are not enough to explain all the missing X-ray reflections. For phenomena on an atomic scale—as X-ray reflection indeed is—finer detail must be considered. For example, if a layer of atoms is laid down, then another laid on top but turned 120° relative to the first, a third laid down but turned 120° relative to the second, and so on, the unit cell must be chosen to include three such layers. The external appearance of such a crystal, interfacial angles, and so forth, will give no evidence of this "screw axis," exhibiting only common trigonal symmetry.

Even before X-rays could be used to test the ideas, Schoenflies and Federov separately saw the possibility of such added finer details and showed that by extending the notion of axes of symmetry to include screw axes and the notion of a plane of symmetry to include cases of the image being displaced parallel to the image plane by, say, a half or quarter cell edge in a regular manner (so-called glide planes), there might be many ways of arranging atoms in unit cells. Schoenflies and Federov mapped these axes, screw axes, mirror planes, and glide planes to find self-consistent arrangements. They disclosed 230 such arrangements that could be fitted into unit cells from which a crystal could be constructed by translation only.

The list of screw axes is as follows:

1. A twofold (binary) screw symbolized as 2_1.
2. A pair of threefold screw axes, 3_1 and 3_2. One makes a right-handed helix, the other a left-handed helix.
3. Three fourfold screw axes, 4_1, 4_2, 4_3. The 4_2 has no handedness, the other two do have handedness.
4. Five sixfold screw axes, 6_1, 6_2, 6_3, 6_4, 6_5. Of these 6_3 has no handedness.

The list of glide planes is as follows:

1. A reflection in a plane parallel to the a axis followed by a translation $a/2$ parallel to the a axis. This is symbolized by the letter a. The letter b refers to a similar reflection in a plane parallel to the b axis followed by a translation $b/2$. The letter c symbolizes a reflection in a plane parallel to the c axis followed by a translation of $c/2$. These three are called axial glides.
2. Diagonal glides have displacement components $a/2 + b/2$ or $b/2 + c/2$ or $c/2 + a/2$. The symbol is n.
3. Diamond glides, symbolized by the letter d, have translations $b/4 + c/4$ or $c/4 + a/4$ or $a/4 + b/4$. Glide planes as well as screw axes cause the elimination of certain X-ray reflections.

It might seem that the letters a, b, c are overworked. At first a, b, and c were lines directed along three intersecting edges of a unit cell. The lengths of these axes a, b, c were proportional to the lengths of the unit cell edges. The unit cell edges themselves were written as a_0, b_0, and c_0, generally given in angstroms. Moreover, A, B, and C were used to designate base-centered cells. We find that one of the 230 space groups has the designation Pc. The large P indicates that the cell is primitive (i.e., not face or body or end centered). The small c tells us that there is a glide plane of translation $c/2$; another space group is designated as $C2/c$. The large C indicates a base-centered cell, the 2 reveals that the cell has a binary axis, and the small c shows that there is an axial glide plane with translation $c/2$ parallel to the c axis. We see that the context keeps clear this multiple use of a, b, c and A, B, C.

X-Rays

The most useful X-ray for crystal orientation is copper Kα, which has a wavelength of about 1.5405 Å. This wavelength is slightly shorter than the distance between atoms in a crystal, and this property makes X-ray diffraction practical. In Fig. 3.1 we see that the rays from a distance source striking the two atom layers and scattering in all directions are "in step" along a departing path, if the distance 1, 2, 3 is an integral multiple of the ray wavelength. Hence all layers scatter "in step" along the departing direction. The distance 1, 2, 3 is $2d \sin \theta$, where d is the distance between layers and θ is the arrival angle, which turns out to be also the departure angle. The departing rays lie in a plane defined by the arriving rays and the normal of the atomic plane. Hence we think of this as reflection of rays by the atomic plane. This reflection differs from ordinary light reflection from a mirror in that θ must have certain values defined by the condition

$$n\lambda = 2d \sin \theta \qquad (3.1)$$

This relation is known as Bragg's law.

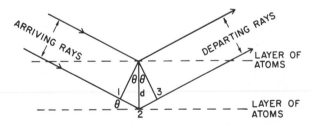

Figure 3.1 X-ray reflection.

The Cone of Reflection

In Fig. 3.2 rays impinging on the crystal at O along the path $1-O$ will be reflected along path $O-1'$; those coming in along path $2-O$ will depart along $O-2'$, and those coming in along path $3-O$ will depart along path $O-3'$. All these paths make an angle $(90° - \theta)$ with the atomic plane normal; hence they define a circular cone of half-angle $(90° - \theta)$. If the crystal surface is not parallel to these atomic planes, the reflection is still from the atomic planes, not from the crystal surface. Thus it is clear that if the crystal is rotated about its atomic plane, normal reflections will not be affected. If the crystal is rotated about some other axis, however, the reflections from this set of atomic planes will be lost, although some other atomic planes may reflect, but along a new direction.

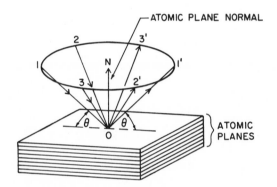

Figure 3.2   Cone of reflection.

If the crystal is rotated about an axis perpendicular both to the atomic normal N and to the direction $1-O$ along which a ray is traveling, a new reflection may be found, a reflection with a different n and a different θ for Eq. 3.1. Not all values of n are valid for a given crystal. For example, the (111) plane of germanium reflects copper $K\alpha$ rays with $\theta = 13°38'$ for $n = 1$. If we substitute $n = 2$ in Eq. 3.1, although we can calculate the angle as $28°8'$, we do not find the reflection. However, $n = 3$ or 4 will give bona fide reflections, but possibly with a different intensity in the reflected beam. If we try $n = 5$ we find that $\sin \theta = 1.1786$, and there is no such angle. Yet if we use shorter wavelength X-rays we can find values of n exceeding 4. Thus we are led to look for the rules governing the possibility of X-ray reflections. The crystallographers who have studied space groups have prepared such a list, which is presented as Table 3.1. Those who have examined the intensities of many X-ray reflections for a given substance have been able to assign the substance to a definite space

TABLE 3.1

Point group		Space group	Conditions limiting possible reflections
		Triclinic	
1		$C_1^1\text{-}P\,1$	None
$\bar{1}$		$C_i^1\text{-}P\,\bar{1}$	None
		Monoclinic	
2		$C_2^1\text{-}P2$	None
2		$C_2^2\text{-}P2_1$	$00l:l=2n$
2		$C_2^3\text{-}B2$	$hkl:h+l=2n$
m		$C_s^1\text{-}Pm$	None
m		$C_s^2\text{-}Pb$	$hk0:k=2n$
m		$C_s^3\text{-}Bm$	$hkl:h+l=2n$
m		$C_s^4\text{-}Bb$	$hkl:h+l=2n,\ hk0:k=2n$
$2/m$		$C_{2h}^1\text{-}P2/m$	None
$2/m$		$C_{2h}^2\text{-}P2_1/m$	$00l:l=2n$
$2/m$		$C_{2h}^3\text{-}B2/m$	$hkl:h+l=2n$
$2/m$		$C_{2h}^4\text{-}P2/b$	$hk0:k=2n$
$2/m$		$C_{2h}^5\text{-}P2_1/b$	$hk0:k=2n,\ 00l:l=2n$
$2/m$		$C_{2h}^6\text{-}B2/b$	$hkl:h+l=2n,\ hk0:k=2n$
		Orthorhombic	
222	V^1,	$D_2^1\text{-}P222$	None
222	V^2,	$D_2^2\text{-}P222_1$	$00l:l=2n$
222	V^3,	$D_2^3\text{-}P2_12_12$	$h00:h=2n,\ 0k0:k=2n$
222	V^4,	$D_2^4\text{-}P2_12_12_1$	$h00:h=2n,\ 0k0:k=2n,\ 00l:l=2n$
222	V^5,	$D_2^5\text{-}C222_1$	$hkl:h+k=2n,\ 00l:l=2n$
222	V^6,	$D_2^6\text{-}C222$	$hkl:h+k=2n$
222	V^7,	$D_2^7\text{-}F222$	$hkl:h+k=2n$ and $k+l=2n$
222	V^8,	$D_2^8\text{-}I222$	$hkl:h+k+l=2n$
222	V^9,	$D_2^9\text{-}I2_12_12$	$hkl:h+k+l=2n$
$2mm$		$C_{2v}^1\text{-}Pmm2$	None
$2mm$		$C_{2v}^2\text{-}Pmc2_1$	$h0l:l=2n$
$2mm$		$C_{2v}^3\text{-}Pcc2$	$0kl:l=2n,\ h0l:l=2n$
$2mm$		$C_{2v}^4\text{-}Pma2$	$h0l:h=2n$
$2mm$		$C_{2v}^5\text{-}Pca2_1$	$0kl:l=2n,\ h0l:h=2n$
$2mm$		$C_{2v}^6\text{-}Pnc2$	$0kl:k+l=2n,\ h0l:l=2n$
$2mm$		$C_{2v}^7\text{-}Pmn2_1$	$h0l:l+h=2n$
$2mm$		$C_{2v}^8\text{-}Pba2$	$0kl:k=2n,\ h0l:h=2n$
$2mm$		$C_{2v}^9\text{-}Pna2_1$	$0kl:k+l=2n,\ h0l:h=2n$
$2mm$		$C_{2v}^{10}\text{-}Pnn2$	$0kl:k+l=2n,\ h0l:l+h=2n$
$2mm$		$C_{2v}^{11}\text{-}Cmm2$	$hkl:h+k=2n$
$2mm$		$C_{2v}^{12}\text{-}Cmc2_1$	$hkl:h+k=2n$
$2mm$		$C_{2v}^{13}\text{-}Ccc2$	$hkl:h+k=2n,\ 0kl:l=2n,\ h0l:l=2n$
$2mm$		$C_{2v}^{14}\text{-}Amm2$	$hkl:k+l=2n$
$2mm$		$C_{2v}^{15}\text{-}Abm2$	$hkl:k+l=2n,\ 0kl:k=2n$
$2mm$		$C_{2v}^{16}\text{-}Ama2$	$hkl:k+l=2n,\ h0l:h=2n$
$2mm$		$C_{2v}^{17}\text{-}Aba2$	$hkl:k+l=2n,\ 0kl:k=2n,\ h0l:h=2n$
$2mm$		$C_{2v}^{18}\text{-}Fmm2$	$hkl:h+k=2n$ and $k+l=2n$
$2mm$		$C_{2v}^{19}\text{-}Fdd2$	$hkl:h+k=2n$ and $k+l=2n,\ 0kl:k+l=4n,\ h0l:l+h=4n$
$2mm$		$C_{2v}^{20}\text{-}Imm2$	$hkl:h+k+l=2n$
$2mm$		$C_{2v}^{21}\text{-}Iba2$	$hkl:h+k+l=2n,\ 0kl:k=2n,\ h0l:h=2n$
$2mm$		$C_{2v}^{22}\text{-}Ima2$	$hkl:h+k+l=2n,\ h0l:h=2n$
mmm	V_h^1,	$D_{2h}^1\text{-}Pmmm$	None

TABLE 3.1 (*Continued*)

Point group	Space group	Conditions limiting possible reflections
mmm V_h^2,	D_{2h}^2-$Pnnn$	$0kl:k+l=2n,\ h0l:l+h=2n,\ hk0:h+k=2n$
mmm V_h^3,	D_{2h}^3-$Pccm$	$0kl:l=2n,\ h0l:l=2n$
mmm V_h^4,	D_{2h}^4-$Pban$	$0kl:k=2n,\ h0l:h=2n,\ hk0:h+k=2n$
mmm V_h^5,	D_{2h}^5-$Pmma$	$hk0:h=2n$
mmm V_h^6,	D_{2h}^6-$Pnna$	$0kl:k+l=2n,\ h0l:l+h=2n,\ hk0:h=2n$
mmm V_h^7,	D_{2h}^7-$Pmna$	$h0l:l+h=2n,\ hk0:h=2n$
mmm V_h^8,	D_{2h}^8-$Pcca$	$0kl:l=2n,\ h0l:l=2n,\ hk0:h=2n$
mmm V_h^9,	D_{2h}^9-$Pbam$	$0kl:k=2n,\ h0l:h=2n$
mmm V_h^{10},	D_{2h}^{10}-$Pccn$	$0kl:l=2n,\ h0l:l=2n,\ hk0:h+k=2n$
mmm V_h^{11},	D_{2h}^{11}-$Pbcm$	$0kl:k=2n,\ h0l:l=2n$
mmm V_h^{12},	D_{2h}^{12}-$Pnnm$	$0kl:k+l=2n,\ h0l:l+h=2n$
mmm V_h^{13},	D_{2h}^{13}-$Pmmn$	$hk0:h+k=2n$
mmm V_h^{14},	D_{2h}^{14}-$Pbcn$	$0kl:k=2n,\ h0l:l=2n,\ hk0:h+k=2n$
mmm V_h^{15},	D_{2h}^{15}-$Pbca$	$0kl:k=2n,\ h0l:l=2n,\ hk0:h=2n$
mmm V_h^{16},	D_{2h}^{16}-$Pnma$	$0kl:k+l=2n,\ hk0:h=2n$
mmm V_h^{17},	D_{2h}^{17}-$Cmcm$	$hkl:h+k=2n,\ h0l:l=2n$
mmm V_h^{18},	D_{2h}^{18}-$Cmca$	$hkl:h+k=2n,\ h0l:l=2n,\ hk0:h=2n$
mmm V_h^{19},	D_{2h}^{19}-$Cmmm$	$hkl:h+k=2n$
mmm V_h^{20},	D_{2h}^{20}-$Cccm$	$hkl:h+k=2n,\ 0kl:l=2n,\ h0l:l=2n$
mmm V_h^{21},	D_{2h}^{21}-$Cmma$	$hkl:h+k=2n$
mmm V_h^{22},	D_{2h}^{22}-$Ccca$	$hkl:h+k=2n,\ 0kl:l=2n,\ h0l:l=2n,\ hk0:h=2n$
mmm V_h^{23},	D_{2h}^{23}-$Fmmm$	$hkl:h+k=2n$ and $h+l=2n$
mmm V_h^{24},	D_{2h}^{24}-$Fddd$	$hkl:h+k=2n$ and $k+l=2n,\ h0l:l+h=4n,\ hk0,\ h+k=4n,$ $0kl:k+l=4n$
mmm V_h^{25},	D_{2h}^{25}-$Immm$	$hkl:h+k+l=2n$
mmm V_h^{26},	D_{2h}^{26}-$Ibam$	$hkl:h+k+l=2n,\ 0kl:k=2n,\ h0l:h=2n$
mmm V_h^{27},	D_{2h}^{27}-$Ibca$	$hkl:h+k+l=2n,\ 0kl:k=2n,\ h0l:l=2n, hk0:h=2n$
mmm V_h^{28},	D_{2h}^{28}-$Imma$	$hkl:h+k+l=2n,\ hk0:h=2n$
	Tetragonal	
4	C_4^1-$P4$	None
4	C_4^2-$P4_1$	$00l:l=4n$
4	C_4^3-$P4_2$	$00l:l=2n$
4	C_4^4-$P4_3$	$00l:l=4n$
4	C_4^5-$I4$	$hkl:h+k+l=2n$
4	C_4^6-$I4_1$	$hkl:h+k+l=2n,\ 00l:l=4n$
$\bar{4}$	S_4^1-$P\bar{4}$	None
$\bar{4}$	S_4^2-$I\bar{4}$	$hkl:h+k+l=2n$
$4/m$	C_{4h}^1-$P4/m$	None
$4/m$	C_{4h}^2-$P4_2/m$	$00l:l=2n$
$4/m$	C_{4h}^3-$P4/n$	$hk0:h+k=2n$
$4/m$	C_{4h}^4-$P4_2/n$	$hk0:h+k=2n,\ 00l:l=2n$
$4/m$	C_{4h}^5-$I4/m$	$hkl:h+k+l=2n$
$4/m$	C_{4h}^6-$I4_1/a$	$hkl:h+k+l=2n,\ hk0:h$ and $k=2n,\ 00l:l=4n$
422	D_4^1-$P422$	None
422	D_4^2-$P42_12$	$h00:h=2n$
422	D_4^3-$P4_122$	$00l:l=4n$
422	D_4^4-$P4_12_12$	$00l:l=4n,\ h00:h=2n$
422	D_4^5-$P4_222$	$00l:l=2n$
422	D_4^6-$P4_22_12$	$00l:l=2n,\ h00:h=2n$
422	D_4^7-$P4_322$	$00l:l=4n$

TABLE 3.1 (*Continued*)

Point group		Space group	Conditions limiting possible reflections
422		D_4^8-$P4_32_12$	$00l: l = 4n$, $h00: h = 2n$
422		D_4^9-$I422$	$hkl: h + k + l = 2n$
422		D_4^{10}-$I4_122$	$hkl: h + k + l = 2n$, $00l: l = 4n$
$4mm$		C_{4v}^1-$P4mm$	None
$4mm$		C_{4v}^2-$P4bm$	$0kl: k = 2n$
$4mm$		C_{4v}^3-$P4_2cm$	$0kl: l = 2n$
$4mm$		C_{4v}^4-$P4_2nm$	$0kl: k + l = 2n$
$4mm$		C_{4v}^5-$P4cc$	$0kl: l = 2n$, $hhl: l = 2n$
$4mm$		C_{4v}^6-$P4nc$	$0kl: k + l = 2n$, $hhl: l = 2n$
$4mm$		C_{4v}^7-$P4_2mc$	$hhl: l = 2n$
$4mm$		C_{4v}^8-$P4_2bc$	$0kl: k = 2n$, $hhl: l = 2n$
$4mm$		C_{4v}^9-$I4mm$	$hkl: h + k + l = 2n$
$4mm$		C_{4v}^{10}-$I4cm$	$hkl: h + k + l = 2n$, $0kl: l = 2n$
$4mm$		C_{4v}^{11}-$I4_1md$	$hkl: h + k + l = 2n$, $hhl: 2h + l = 4n$
$4mm$		C_{4v}^{12}-$I4_1cd$	$hkl: h + k + l = 2n$, $0kl: l = 2n$, $hhl: 2h + l = 4n$
$\bar{4}2m$	V_d^1,	D_{2d}^1-$P\bar{4}2m$	None
$\bar{4}2m$	V_d^2,	D_{2d}^2-$P\bar{4}2c$	$hhl: l = 2n$
$\bar{4}2m$	V_d^3,	D_{2d}^3-$P\bar{4}2_1m$	$h00: h = 2n$
$\bar{4}2m$	V_d^4,	D_{2d}^4-$P\bar{4}2_1c$	$hhl: l = 2n$, $h00: h = 2n$
$\bar{4}2m$	V_d^5,	D_{2d}^5-$P\bar{4}m2$	None
$\bar{4}2m$	V_d^6,	D_{2d}^6-$P\bar{4}c2$	$0kl: l = 2n$
$\bar{4}2m$	V_d^7,	D_{2d}^7-$P\bar{4}b2$	$0kl: k = 2n$
$\bar{4}2m$	V_d^8,	D_{2d}^8-$P\bar{4}n2$	$0kl: k + l = 2n$
$\bar{4}2m$	V_d^9,	D_{2d}^9-$I\bar{4}m2$	$hkl: h + k + l = 2n$
$\bar{4}2m$	V_d^{10},	D_{2d}^{10}-$I\bar{4}c2$	$hkl: h + k + l = 2n$, $0kl: l = 2n$
$\bar{4}2m$	V_d^{11},	D_{2d}^{11}-$I\bar{4}2m$	$hkl: h + k + l = 2n$
$\bar{4}2m$	V_d^{12},	D_{2d}^{12}-$I\bar{4}2d$	$hkl: h + k + l = 2n$, $hhl: 2h + l = 4n$
$4/mmm$		D_{4h}^1-$P4/mmm$	None
$4/mmm$		D_{4h}^2-$P4/mcc$	$0kl: l = 2n$, $hhl: l = 2n$
$4/mmm$		D_{4h}^3-$P4/nbm$	$hk0: h + k = 2n$, $0kl: k = 2n$
$4/mmm$		D_{4h}^4-$P4/nnc$	$hk0: h + k = 2n$, $0kl: k + l = 2n$, $hhl: l = 2n$
$4/mmm$		D_{4h}^5-$P4/mbm$	$0kl: k = 2n$
$4/mmm$		D_{4h}^6-$P4/mnc$	$0kl: k + l = 2n$, $hhl: l = 2n$
$4/mmm$		D_{4h}^7-$P4/nmm$	$hk0: h + k = 2n$
$4/mmm$		D_{4h}^8-$P4/ncc$	$hk0: h + k = 2n$, $0kl: l = 2n$, $hhl: l = 2n$
$4/mmm$		D_{4h}^9-$P4_2/mmc$	$hhl: l = 2n$
$4/mmm$		D_{4h}^{10}-$P4_2/mcm$	$0kl: l = 2n$
$4/mmm$		D_{4h}^{11}-$P4_2/nbc$	$hk0: h + k = 2n$, $0kl: k = 2n$, $hhl: l = 2n$
$4/mmm$		D_{4h}^{12}-$P4_2/nnm$	$hk0: h + k = 2n$, $0kl: k + l = 2n$
$4/mmm$		D_{4h}^{13}-$P4_2/mbc$	$0kl: k = 2n$, $hkl: l = 2n$
$4/mmm$		D_{4h}^{14}-$P4_2/mnm$	$0kl: k + l = 2n$
$4/mmm$		D_{4h}^{15}-$P4_2/nmc$	$hk0: h + k = 2n$, $hhl: l = 2n$
$4/mmm$		D_{4h}^{16}-$P4_2/ncm$	$hk0: h + k = 2n$, $0kl: l = 2n$
$4/mmm$		D_{4h}^{17}-$I4/mmm$	$hkl: h + k + l = 2n$
$4/mmm$		D_{4h}^{18}-$I4/mcm$	$hkl: h + k + l = 2n$, $0kl: l = 2n$
$4/mmm$		D_{4h}^{19}-$I4_1/amd$	$hkl: h + k + l = 2n$, $hk0: h = 2n$, $hhl: 2h + l = 4n$
$4/mmm$		D_{4h}^{20}-$I4_1/acd$	$hkl: h + k + l = 2n$, $hk0: h = 2n$, $0kl: l = 2n$, $hhl: 2h + l = 4n$
		Trigonal	
3		C_3^1-$P3$	None
3		C_3^2-$P3_1$	$00 \cdot l: l = 3n$

TABLE 3.1 (*Continued*)

Point group		Space group	Conditions limiting possible reflections
3		C_3^3-$P3_2$	$00\cdot l: l = 3n$
3		C_3^4-$R3$	$hk\cdot l: -h + k + l = 3n$
$\bar{3}$	$S_6^1,$	C_{3i}^1-$P\bar{3}$	None
$\bar{3}$	$S_6^2,$	C_{3i}^2-$R\bar{3}$	$hk\cdot l: -h + k + 1 = 3n$
32		D_3^1-$P312$	None
32		D_3^2-$P321$	None
32		D_3^3-$P3_112$	$00\cdot l: l = 3n$
32		D_3^4-$P3_121$	$00\cdot l: l = 3n$
32		D_3^5-$P3_212$	$00\cdot l: l = 3n$
32		D_3^6-$P3_221$	$00\cdot l: l = 3n$
32		D_3^7-$R32$	$hk\cdot l: -h + k + l = 3n$
$3m$		C_{3v}^1-$P3m1$	None
$3m$		C_{3v}^2-$P31m$	None
$3m$		C_{3v}^3-$P3c1$	$h\bar{h}\cdot l: l = 2n$
$3m$		C_{3v}^4-$P31c$	$hh\cdot l: l = 2n$
$3m$		C_{3v}^5-$R3m$	$hk\cdot l: -h + k + l = 3n$
$3m$		C_{3v}^6-$R3c$	$hk\cdot l: -h + k + l = 3n,\ h\bar{h}\cdot l: l = 2n$
$\bar{3}m$		D_{3d}^1-$P\bar{3}1m$	None
$\bar{3}m$		D_{3d}^2-$P\bar{3}1c$	$hh\cdot l: l = 2n$
$\bar{3}m$		D_{3d}^3-$P\bar{3}m1$	None
$\bar{3}m$		D_{3d}^4-$P\bar{3}c1$	$h\bar{h}\cdot l: l = 2n$
$\bar{3}m$		D_{3d}^5-$R\bar{3}m$	$hk\cdot l: -h + k + l = 3n$
$\bar{3}m$		D_{3d}^6-$R\bar{3}c$	$hk\cdot l: -h + k + l = 3n,\ h\bar{h}\cdot l: l = 2n$
		Hexagonal	
6		C_6^1-$P6$	None
6		C_6^2-$P6_1$	$00\cdot l: l = 6n$
6		C_6^3-$P6_5$	$00\cdot l: l = 6n$
6		C_6^4-$P6_2$	$00\cdot l: l = 3n$
6		C_6^5-$P6_4$	$00\cdot l: l = 3n$
6		C_6^6-$P6_3$	$00\cdot l: l = 2n$
$\bar{6}$		C_{3h}^1-$P\bar{6}$	None
$6/m$		C_{6h}^1-$P6/m$	None
$6/m$		C_{6h}^2-$P6_3/m$	$00\cdot l: l = 2n$
622		D_6^1-$P622$	None
622		D_6^2-$P6_122$	$00\cdot l: l = 6n$
622		D_6^3-$P6_522$	$00\cdot l: l = 6n$
622		D_6^4-$P6_222$	$00\cdot l: l = 3n$
622		D_6^5-$P6_422$	$00\cdot l: l = 3n$
622		D_6^6-$P6_322$	$00\cdot l: l = 2n$
$6mm$		C_{6v}^1-$P6mm$	None
$6mm$		C_{6v}^2-$P6cc$	$hh\cdot l: l = 2n,\ h\bar{h}\cdot l: l = 2n$
$6mm$		C_{6v}^3-$P6_3cm$	$h\bar{h}\cdot l: l = 2n$
$6mm$		C_{6v}^4-$P6_3mc$	$hh\cdot l: l = 2n$
$\bar{6}m2$		D_{3h}^1-$P\bar{6}m2$	None
$\bar{6}m2$		D_{3h}^2-$P\bar{6}c2$	$h\bar{h}\cdot l: l = 2n$
$\bar{6}m2$		D_{3h}^3-$P\bar{6}2m$	None
$\bar{6}m2$		D_{3h}^4-$P\bar{6}2c$	$hh\cdot l: l = 2n$
622		D_{6h}^1-$P6/mmm$	None
622		D_{6h}^2-$P6/mcc$	$hh\cdot l: l = 2n,\ h\bar{h}\cdot l: l = 2n$
622		D_{6h}^3-$P6_3/mcm$	$h\bar{h}\cdot l: l = 2n$

TABLE 3.1 *(Continued)*

Point group	Space group	Conditions limiting possible reflections
622	D_{6h}^4-$P6_3/mmc$	$hh\cdot l:l = 2n$
	Cubic	
23	T^1-$P23$	None
23	T^2-$F23$	$hkl:h + k$ and $k + l = 2n$
23	T^3-$I23$	$hkl:h + k + l = 2n$
23	T^4-$P2_13$	$h00:h = 2n$, permuted
23	T^5-$I2_13$	$hkl:h + k + l = 2n$
$m3$	T_h^1-$Pm3$	None
$m3$	T_h^2-$Pn3$	$0kl:k + l = 2n$, permuted
$m3$	T_h^3-$Fm3$	$hkl:h + k$ and $k + l = 2n$
$m3$	T_h^4-$Fd3$	$hkl:h + k$ and $k + l = 2n$, $0kl:k + l = 4n$, permuted
$m3$	T_h^5-$Im3$	$hkl:h + k + l = 2n$
$m3$	T_h^6-$Pa3$	$0kl:k = 2n$, h, k, l, not permutable
$m3$	T_h^7-$Ia3$	$hkl:h + k + l = 2n$, $0kl:k = 2n$, permuted
432	O^1-$P432$	None
432	O^2-$P4_232$	$h00:h = 2n$, permuted
432	O^3-$F432$	$hkl:h + k +$ and $k + l = 2n$, permuted
432	O^4-$F4_132$	$hkl:h + k$ and $k + l = 2n$, $h00:h = 4n$, permuted
432	O^5-$I432$	$hkl:h + k + l = 2n$
432	O^6-$P4_332$	$h00:h = 4n$, permuted
432	O^7-$P4_132$	$h00:h = 4n$, permuted
432	O^8-$I4_132$	$hkl:h + k + l = 2n$, $h00:h = 4n$, permuted
$\bar{4}3m$	T_d^1-$P\bar{4}3m$	None
$\bar{4}3m$	T_d^2-$F\bar{4}3m$	$hkl:h + k$ and $k + l = 2n$
$\bar{4}3m$	T_d^3-$I\bar{4}3m$	$hkl:h + k + l = 2n$
$\bar{4}3m$	T_d^4-$P\bar{4}3n$	$hhl:l = 2n$, permuted
$\bar{4}3m$	T_d^5-$F\bar{4}3c$	$hkl:h + k$ and $k + l = 2n$, $hhl:l = 2n$, permuted
$\bar{4}3m$	T_d^6-$I\bar{4}3d$	$hkl:h + k + l = 2n$, $hhl:2h + l = 4n$, permuted
$m3m$	O_h^1-$Pm3m$	None
$m3m$	O_h^2-$Pn3n$	$hhl:l = 2n$, permuted. $0kl:k + l = 2n$, permuted
$m3m$	O_h^3-$Pm3n$	$hhl:l = 2n$, permuted
$m3m$	O_h^4-$Pn3m$	$0kl:k + l = 2n$, permuted
$m3m$	O_h^5-$Fm3m$	$hkl:h + k$ and $k + l = 2n$
$m3m$	O_h^6-$Fm3c$	$hkl:h + k$ and $k + l = 2n$, $hhl:l = 2n$, permuted
$m3m$	O_h^7-$Fdm3$	$hkl:h + k$ and $k + l = 2n$, $0kl:k + l = 4n$, permuted
$m3m$	O_h^8-$Fd3c$	$hkl:h + k$ and $k + l = 2n$, $hhl:l = 2n$, permuted, $0kl:k + l = 4n$, permuted
$m3m$	O_h^9-$Im3m$	$hkl:h + k + l = 2n$
$m3m$	O_h^{10}-$Ia3d$	$hkl:h + k + l = 2n$, $hhl:2h + l = 4n$, permuted

group. We need this information in using X-rays for orientation. The double notation (Schoenflies and Hermann-Mauguin) is used because some of the older sources of information on materials use Schoenflies notation only. Even the Schoenflies notation has variants. The D_2^n's often appear as V^n. The D_{2h}^n's then appear as V_h^n, the D_{2d}^n's appear as V_d^n, and finally C_{3i}^1 and C_{3i}^2 sometimes appear as S_6^1 and S_6^2. We give both versions in Table 3.1.

As an example we consider the material lithium iodate, $LiIO_3$. The fortieth edition of the *Handbook of Chemistry and Physics* gives this as hexagonal, D_6^6 with $a_0 = 5.469$ Å, $c_0 = 5.155$ Å with two molecules of $LiIO_3$ in each unit cell (p. 2694). Table 3.1 gives this as $P6_322$ in point group 662 and says that the only restriction on X-ray reflections is that in planes of the type $00 \cdot l$, l must be even.

A note of caution is in order. It is well to check such data with other information. On page 2692 of the fortieth edition we find potassium dihydrogen phosphate KH_2PO_4, commonly called KDP, as tetragonal, space group V_h^{12}, with $a_0 = 7.43$ Å, $c_0 = 6.97$ Å. This is inconsistent because V_h^{12} is an orthorhombic group, not a tetragonal group. Moreover, V_h^{12} is of point group mmm, which has no piezoactivity, and KDP is very active piezoelectrically. Also KDP is optically uniaxial, as we discuss later, and tetragonal crystals should be uniaxial, whereas orthorhombic crystals are biaxial. The files of the American Society for Testing and Materials (ASTM) contain a card for KDP listing that substance as $D_{2d}^{12}\text{-}I\bar{4}2d$, with $a_0 = 7.448$ Å, $c_0 = 6.977$ Å, 4 molecules per unit cell. Table 3.1 gives the conditions for X-ray reflection as $hkl: h + k + l = 2n$, $hhl: 2h + l = 4n$. This means that in general, $h + k + l$ must be even, but in case $k = h$, then $h + k + l$ must be divisible by 4. The ASTM card for KDP gives the d spacing in angstroms for 36 reflections and also gives the intensities of these lines. This can be very helpful, but Bragg angles θ calculated from the larger d spacings may not be very accurate. The fortieth edition of *The Handbook of Chemistry and Physics* lists $LiCbO_3$ as belonging to space group C_{3i}^2 with $a = 5.47$, $\alpha = 55°43'$. This is a nonpiezoactive space group. However columbium (Cb) is the old name for niobium. The ASTM cards list lithium niobium oxide ($LiNbO_3$) as belonging to space group $R3C$ with $a_0 = 5.1494$ Å, $c_0 = 13.8620$ Å. This is a piezoactive space group, and $LiNbO_3$ is very piezoactive. Using the given a_0 and c_0, we calculate the rhombohedral lattice constants as $a = 5.49$ Å, $\alpha = 55°53'$ (see Eqs. 2.1' and 2.2').

d Spacings

For any crystal the d spacings decrease as $h, k,$ and l increase. For cubic crystals the relation is quite simple:

$$\frac{1}{d(hkl)} = \frac{1}{a_0} \sqrt{h^2 + k^2 + l^2} \tag{3.2}$$

We give $1/d$ instead of d because

$$\sin \theta = \frac{\lambda}{2} \times \frac{1}{d} \tag{3.3}$$

It has been found very convenient to drop the n in Bragg's equation by absorbing it into the $d : d(nh, nk, nl) = d(hkl)/n$. This is a great help in plotting reflections. It also follows directly from Eq. 3.2. For crystals with hexagonal indexing we have

$$\frac{1}{d(hk \cdot l)} = \sqrt{\frac{4}{3a_0^2}(h^2 + hk + k^2) + \frac{l^2}{c_0^2}} \tag{3.4}$$

For tetragonal crystals we have

$$\frac{1}{d(hkl)} = \sqrt{\frac{h^2 + k^2}{a_0^2} + \frac{l^2}{c_0^2}} \tag{3.5}$$

For orthorhombic crystals we have

$$\frac{1}{d(hkl)} = \sqrt{\frac{h^2}{a_0^2} + \frac{k^2}{b_0^2} + \frac{l^2}{c_0^2}} \tag{3.6}$$

For monoclinic crystals we have

$$\frac{1}{d(hkl)} = \sqrt{\left(\frac{h^2}{a_0^2} + \frac{l^2}{c_0^2} - \frac{2hl \cos \beta}{a_0 c_0}\right)\Big/ \sin^2 \beta + \frac{k^2}{b_0^2}}$$

For triclinic crystals, we let

$$U = \cos \gamma - \cos \alpha \cos \beta$$

$$S = \sqrt{\sin^2 \alpha \sin^2 \beta - U^2}$$

$$D_1 = \frac{h}{a_0 \sin \beta} - \frac{l \cot \beta}{c_0}$$

$$D_2 = \frac{-hU}{a_0 S \sin \beta} + \frac{k \sin \beta}{b_0 S} + \frac{U - \cot \beta - \cos \alpha \sin \beta}{c_0 S}$$

$$D_3 = \frac{l}{c_0}$$

$$\frac{1}{d} = \sqrt{D_1^2 + D_2^2 + D_3^2}$$

GRAPHIC AIDS FOR CUBIC AND UNIAXIAL CRYSTALS

For primitive cubic crystals P, body-centered cubic crystals I, face-centered cubic crystals F, diamond structured crystals D, and spinel structures S, the permitted reflections can be read from Fig. 3.3. If the edge of a strip of paper is marked with a reference mark at (100) and the permitted reflections are noted all along the edge and the strip transferred

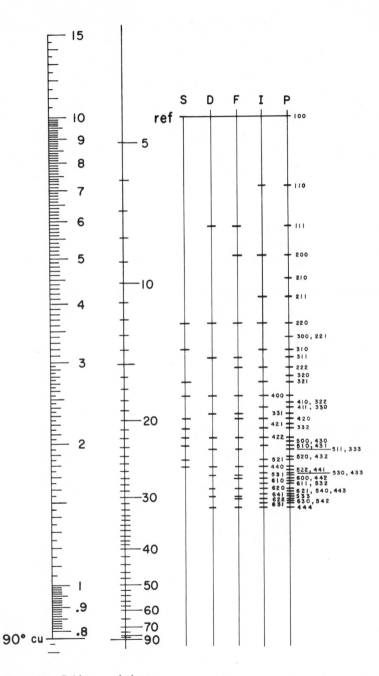

Figure 3.3 Cubic crystal chart.

to the d scale with the (100) reference placed at the reading for the proper lattice constant a_0, the d spacings may be read off directly. If, however, we mark the point "Cu 90°" and transfer to the θ scale with this last mark at the point $\theta = 90°$, we can read all the θ's directly. When a cubic crystal is indexed, its lattice constant and type can be stated.

Using the chart of Fig. 3.3, we can "index" a cubic crystal. That is, if a number of d spacings are known, they can be marked on the edge of a strip of paper employing the left scale of Fig. 3.3. If these markings can be matched to one of the scales to the right, the crystal is cubic and the different h, k, l's can be identified. The easiest way of obtaining such a list of d spacings for an unknown crystal is the powder method.

Powder Methods

The powder camera illustrated in Fig. 3.4 is the simplest equipment that will give a list of d spacings. The crystal material is ground to a fine powder and stuck to a fine fiber placed at the center of a strip of photographic film curved into a cylinder. A collimator (left) permits a

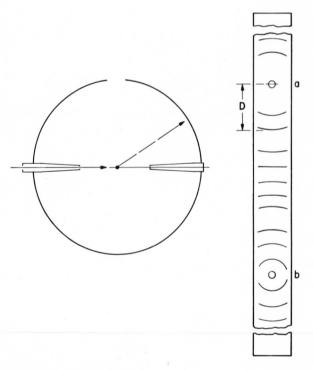

Figure 3.4 Powder camera.

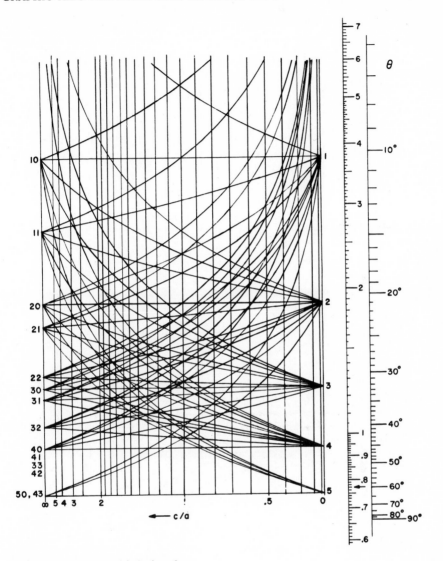

Figure 3.5 Tetragonal indexing chart.

small X-ray beam to strike the powder. Assuming that all orientations are present in the sample, all beams possible will reflect and strike the photographic film. A series of bands appears on the film when it is developed, and the distance along the film from the exit hole is a measure of 2θ. Hence a set of linear measurements D is readily converted into a set of d spacings. This method is commonly used for the identification of

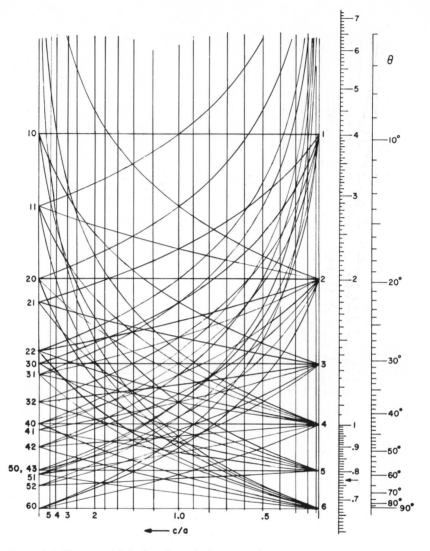

Figure 3.6 Tetragonal indexing chart—body centered.

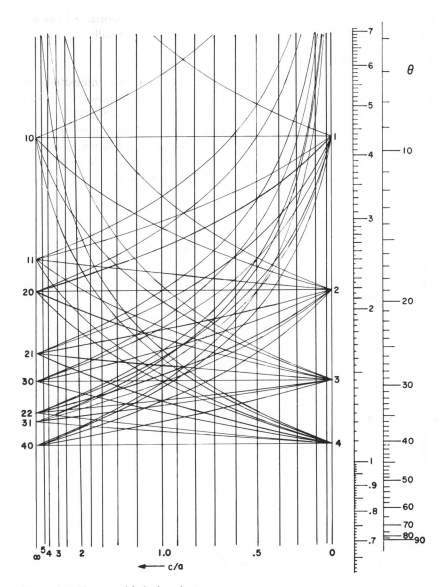

Figure 3.7 Hexagonal indexing chart.

39

materials by comparing the d spacings of the three strongest lines to a catalog of such data. The intensities are estimated visually.

The Powder Diffractometer

To avoid the limitations of visual estimation of line intensities, the diffractometer has been developed. A larger amount of powder is spread over a square centimeter or so area, and reflections are detected by means of a slowly moving radiation detector and recorded on a moving strip of paper. The record is made by a moving pen whose deflection is approximately proportional to the intensity of reflection. The paper is moved so that its advancement is proportional to the motion of the radiation

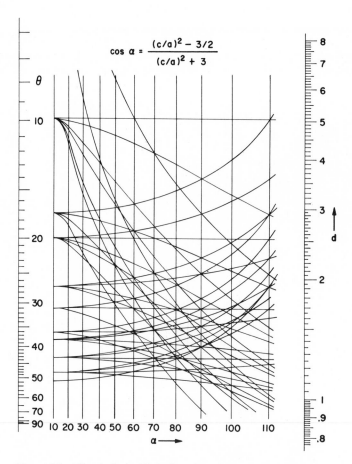

$$\cos \alpha = \frac{(c/a)^2 - 3/2}{(c/a)^2 + 3}$$

Figure 3.8   Rhombohedral indexing chart, hexagonally indexed.

detector. Hence intensity is recorded as a function of 2θ. The data can then be indexed if the crystal symmetry is not too low. Cubic crystals can be immediately indexed by means of the chart (Fig. 3.3).

INDEXING NONCUBIC CRYSTALS

Triclinic, monoclinic, and orthorhombic crystals are best examined by single crystal methods. Here the spatial arrangement of planes with observed Bragg angles is recorded. In the powder method spatial arrangement can only be inferred later. However with tetragonal and hexagonal, i.e. uniaxial crystals, powder data can be indexed.

Indexing uniaxial crystals involves one more variable than cubic crystals require. The unit cell of cubic crystals are equidimensional whereas tetragonal and hexagonal unit cells involve a length-to-width ratio (i.e., a c/a ratio). This is taken care of by graphing the logarithms of d spacings as a function of the c/a ratio. Since these charts become congested, it was thought best to separate the body-centered (Fig. 3.6) from the primitive tetragonal charts (Fig. 3.5) and the rhombohedral from the hexagonal charts (Figs. 3.8 and 3.7, respectively).

Figure 3.5 is a chart for primitive tetragonal crystals, Fig. 3.6 covers body-centered tetragonal crystals. Face-centered tetragonal crystals are best handled by choosing new "a" axes normal to the c axis but 45° to the old "a" axes. This gives a unit cell half as large as the face-centered cell. The new cell is body centered.

Hexagonal crystals and hexagonally indexed rhombohedral crystals are represented by Figs. 3.7 and 3.8, respectively.

4

The X-Ray Orientation Goniometer

Most X-ray orientation goniometers have the basic design presented in Fig. 4.1. Here BW is a beryllium window through which X-rays emerge from the X-ray tube. A collimator permits only a narrow ribbon of rays to strike the crystal XTL. The collimator is a metal tube with narrow slits at the ends. The crystal is secured to a sliding fixture SF that allows the exposed surface to be centered on the axis of rotation AR. The crystal can be swung about this axis and the angle A can be read from a circular scale. For a parallel-sided crystal with a proper atomic plane parallel to the major surfaces, if A is set at the Bragg angle θ, rays reflect at angle 2θ and enter the detector D. The strength of the reflected beam is indicated by the rate meter RM.

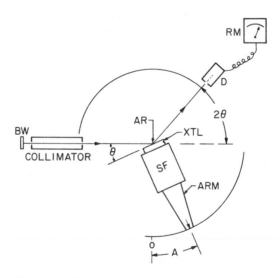

Figure 4.1 The basic crystal orientation goniometer.

If the atomic plane is not parallel to the surface but departs from it by a small clockwise rotation δ about the axis AR, the rate meter will show a maximum at a reading $A = \theta + \delta$. If we turn the crystal end for end, the rate meter will maximize at an angular reading $\bar{A} = \theta - \delta$. Obviously $\delta = (A - \bar{A})/2$. This gives the bases for making rotated cuts and for checking ones already made.

In X-ray checking singly rotated uniaxial plates it is not always obvious where in the plate lies the axis of rotation corresponding to AR of Fig. 4.1. This axis can be found by a systematic search as follows: swing the arm slowly while watching the rate meter, turn the plate a few degrees in its own plane, and swing the arm again, repeating to find the maximum and minimum values of A. The search can be made easier for transparent plates by finding extinction positions between crossed polarizers, as in Fig. 4.2. Turning the plate around as it lies on the polarizer, one sees it brighten up, then darken, then brighten. It becomes dark in four positions. Setting it for the darkest in one of these positions, one draws on the crystal with a felt tip pen a line that is parallel to one of the polarizer axes. Axis AR is either parallel to this line or perpendicular to it. We make an arrowhead on one end of the line just drawn. This helps in interpreting data we will take on the X-ray goniometer.

Assume now that the plate is thick enough to see through the edges (say a millimeter or more thick). Looking along the length, then along the width directions, we see that for one of these directions of viewing we

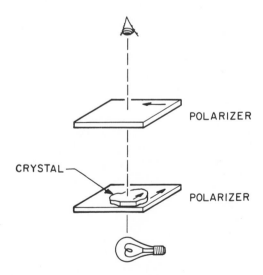

Figure 4.2 Using polarizing filters to mark extinction positions of a crystal plate.

have parallel extinction, for the other, nonparallel extinction. That is, for the greatest darkness for one direction of viewing the plate is parallel to a polarizer axis, for the other direction it is not parallel to a polarizer axis. The viewing direction for "nonparallel extinction" is the axis of rotation AR, and the inclination angle is P or else it is $90° - P$.

In Fig. 4.3 the edge of such a crystal appears sharp at one end to represent the arrow just drawn. This is a plate rotated through an angle P

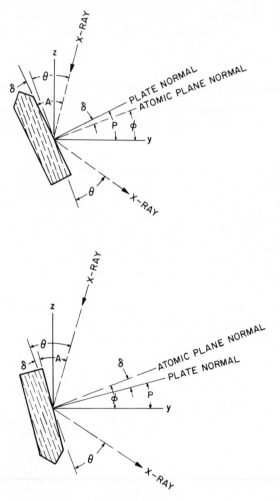

Figure 4.3 Reflection angles: A observed with "arrow up" (top) and "arrow down" (bottom).

counterclockwise from an imagined original position parallel to the $Z–X$ plane. With the arrowhead up (upper figures) reflection occurs for

$$A = \theta + \delta \tag{4.1}$$

With the arrow pointing down (lower figure), we have

$$\bar{A} = \theta - \delta \tag{4.2}$$

whence

$$\delta = \frac{A - \bar{A}}{2} \tag{4.3}$$

From the upper part of Fig. 4.3 we find

$$P = \phi + \delta \tag{4.4}$$

where ϕ is the angle between the atomic plane normal and the y axis.

This gives the value of P on the assumption that the rotation was strictly about the x axis and in the sense intended. If δ is reversed the foregoing analysis will not detect it. To check the sense of the rotation we employ a second reflection, one from an entirely different atomic plane whose normal is also in the $Y–Z$ plane. For this plane ϕ, δ, and θ will be different from the values found for the first plane, but Eqs. 4.1 to 4.4 will apply as

$$A' = \theta' + \delta' \tag{4.1'}$$

$$\bar{A}' = \theta' - \delta' \tag{4.2'}$$

$$\delta' = \frac{A' - \bar{A}'}{2} \tag{4.3'}$$

$$P = \phi' + \delta' \tag{4.4'}$$

If δ has the wrong sign in Eq. 4.1, the value of P from Eq. 4.4 will disagree with the result from Eq. 4.4'.

We choose lithium niobate, a rhombohedral crystal, class $R3C$ for an example. Table 3.1 gives for this class the condition for reflection as $-h + k + l = 3n$ in general; but for planes of the type $h\bar{h} \cdot l$, l must be even. Hence in $00 \cdot l$, $h0 \cdot l$ and $0k \cdot l$, l must be even. For rotated y cuts we need the $(0k \cdot l)$ planes. The normals of all planes $(hk \cdot l)$ of a hexagonal crystal are given by

$$\begin{pmatrix} x \\ y \\ z \end{pmatrix} = \begin{pmatrix} \dfrac{h}{a_0} \\ \dfrac{h + 2k}{a_0\sqrt{3}} \\ \dfrac{l}{c_0} \end{pmatrix} \tag{4.5}$$

For $(0k \cdot l)$ planes, $h = 0$. Hence the angle ϕ between the y axis and the normal to the $(0k \cdot l)$ plane satisfies

$$\tan \phi = \frac{l/c_0}{2k/a_0\sqrt{3}} = \frac{a_0\sqrt{3}}{c_0} \frac{l}{2} \frac{l}{k} \tag{4.6}$$

For any desired rotation angle P we find

$$l = \frac{c_0}{a_0} \times \frac{2k}{\sqrt{3}} \tan P \tag{4.6'}$$

Equation 4.6' may be solved with a slide rule, accepting only $0k \cdot l$'s that are permitted. For $P = 36°$ this gives us $l = 2.26k$, which means that one of the permitted reflections $01 \cdot 2, 02 \cdot 4, \ldots$, may suffice. However it is good to have a "map" of $(0k \cdot l)$ reflections; a plot of

$$\begin{pmatrix} y \\ z \end{pmatrix} = \begin{pmatrix} \dfrac{2k}{\sqrt{3}a_0} \\ \dfrac{l}{c_0} \end{pmatrix}$$

will be such a map.

There is an advantage to multiplying Eq. 4.5 by half the wavelength of the X-ray. This places the points k, l in the direction of the atomic plane normals but also makes the distance of each point from the origin equal to the sine of the respective Bragg angle. A unit circle encloses all the points that can reflect the X-ray wavelength employed. Figure 4.4 is such a map for lithium niobate. Here $a_0 = 5.1494$ Å, $c_0 = 13.8620$ Å.

To X-ray check a 36° Y-rotated plate, we have a choice, according to the map, of $(01 \cdot 2)$, $(02 \cdot 4)$, $(03 \cdot 6)$, or $(04 \cdot 8)$ for the first plane. We pick $(02 \cdot 4)$ and calculate from Eq. 4.5: $\sin \theta = \sqrt{x^2 + y^2 + z^2}$; that is,

$$\sin \theta = \frac{\lambda}{2} \sqrt{\frac{4}{3} \frac{k^2}{a_0^2} + \frac{l^2}{c_0^2}} = 0.41077, \qquad \theta = 24°15'$$

where we have assumed $\lambda = 1.5405$ for copper Kα radiation. Also we have $\tan \phi = z/y$; that is,

$$\tan \phi = \frac{l/c_0}{2k/(a_0\sqrt{3})} = \frac{a_0 l \sqrt{3}}{2c_0 k} = 0.64342, \qquad \phi = 32°45'$$

Since for the second plane δ will be larger than before, it is wise to pick a larger θ. We try $04 \cdot 2$. Calculating as before:

$$\sin \theta' = \frac{\lambda}{2} \sqrt{\frac{4}{3} \times \frac{16}{5.1494^2} + \frac{4}{13.8620^2}} = 0.69976, \qquad \theta' = 44°24'$$

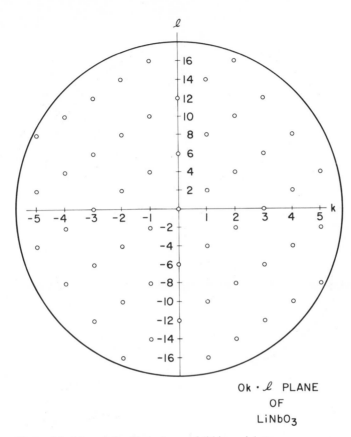

Figure 4.4 Map of the $0k \cdot l$ planes of lithium niobate.

$$\tan \phi' = \frac{5.1494\sqrt{3}}{4 \times 13.862} = 0.16086 \quad \text{or} \quad \phi' = 9°8'$$

In Fig. 4.5 we see the edge of the plate on an $(0k \cdot l)$ map. For this plate δ and δ' are obviously positive. Equation 4.4 gives $\delta = 36° - 32°45' = 3°15'$ and $\delta' = 36° - 9°8' = 26°52'$. Equations 4.1 and 4.2 give $A = 27°30'$, $\bar{A} = 21°0'$, $A' = 71°16'$, $\bar{A}' = 17°32'$.

A certain LiNbO₃ crystal was examined between crossed polarizers and the arrow drawn accordingly. The measurements were as follows:

$h\,k\,l$	Arrow	θ	Arm
02·4	Up	24°15′	19°21′
02·4	Down	24°15′	29°14′
02·4	Right	24°15′	24°55′
02·4	Left	24°15′	23°50′

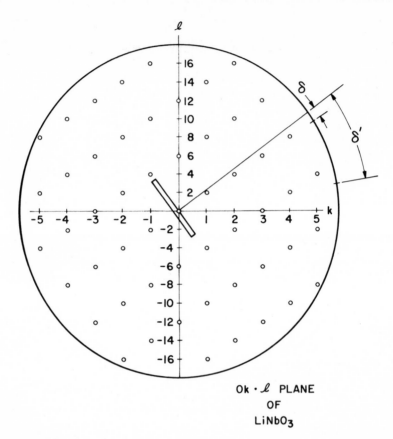

Ok · l PLANE
OF
LiNbO₃

Figure 4.5 Edge view of an oriented crystal of lithium niobate drawn on an $0k \cdot l$ map.

Obviously "up" is the reverse position because the arm reading was smallest for this position. Hence $A = 29°14'$, $\bar{A} = 19°21'$; thus $\delta = 4°56'$ and $P = 37°41'$ by Eq. 4.4. But these measurements do not decide the true sign of δ. Measurements with the detector set for $\theta = 44°24'$ gave

$h\,k\,l$	Arrow	θ	Arm
04·2	Up	44°24′	15°50′
04·2	Down	44°24′	72°57′

whence $\delta = 28°33'$ and $P = 37°41'$.

The right and left readings indicate that the rotation was not strictly about the X axis but about an axis $(24°55' - 23°50')/2 = 32.5$ minutes from the X axis. We learn later that this falsifies the arm readings by about 4 seconds of arc.

Consider now a LiNbO$_3$, Y-rotated plate with $P = 29°30'$. Here $\delta = -3°15'$ if we use the 02·4 plane. Hence arm readings of 27°30' and 21°0' would result, and these seem to agree with those expected for $P = 36°$. However on the (04·2) plane we find arm readings of 64°46' and 24°2', quite different from the expected 17°32' and 71°16'.

With single rotation plates of very uncertain rotation angle P, we can resort to the following subterfuge. Several Bragg angles $\theta_1, \theta_2, \ldots$, and their corresponding arm readings $A_1, A_2, \ldots$, are made and recorded. We plot the polar angle $(A_i - \theta_i)$ against radius vector $\sin \theta_i$ using such a scale that $\sin \theta_i = \sin 90°$ gives a unit circle the same size as that used in making the map (Fig. 4.4). We try to superimpose the plot on the map so that each plotted point $(A_i - \theta_i)$, $\sin \theta_i$ falls on a mapped reflection. It may be necessary to turn the plot over and view through the back. This would be the case if we guessed wrong about the sign of δ—that is, if we were mistaken about which end should carry the arrowhead in Fig. 4.2.

For completeness we now list the rectangular coordinates for the atomic plane normals of the six crystal systems.

Cubic:
$$\begin{pmatrix} h \\ k \\ l \end{pmatrix} \frac{\lambda}{2a_0} \tag{4.7}$$

Hexagonal:
$$\begin{pmatrix} \dfrac{h}{a_0} \\ \dfrac{h+2k}{a_0\sqrt{3}} \\ \dfrac{l}{c_0} \end{pmatrix} \frac{\lambda}{2} \tag{4.8}$$

Tetragonal:
$$\begin{pmatrix} \dfrac{h}{a_0} \\ \dfrac{k}{a_0} \\ \dfrac{l}{c_0} \end{pmatrix} \frac{\lambda}{2} \tag{4.9}$$

Orthorhombic:
$$\begin{pmatrix} \dfrac{h}{a_0} \\ \dfrac{k}{b_0} \\ \dfrac{l}{c_0} \end{pmatrix} \frac{\lambda}{2} \tag{4.10}$$

Monoclinic:
$$\begin{pmatrix} \dfrac{h}{a_0 \sin \beta} - \dfrac{l \cot \beta}{c_0} \\[2mm] \dfrac{k}{b_0} \\[2mm] \dfrac{l}{c_0} \end{pmatrix} \dfrac{\lambda}{2} \qquad (4.11)$$

Triclinic:
$$\begin{pmatrix} \dfrac{h}{a_0 \sin \beta} - \dfrac{l \cot \beta}{c_0} \\[2mm] \dfrac{-hv_1}{a_0 v_2 \sin \beta} + \dfrac{k}{v_2 b_0} + l\,\dfrac{v_1 \cot \beta - \cos \alpha}{v_2 c} \\[2mm] \dfrac{l}{c_0} \end{pmatrix} \dfrac{\lambda}{2} \qquad (4.12)$$

where

$$v_1 = \frac{\cos \gamma - \cos \alpha \cos \beta}{\sin \beta} \qquad (4.13)$$

$$v_2 = \frac{\{1 + 2 \cos \alpha \cos \beta \cos \gamma - (\cos^2 \alpha + \cos^2 \beta + \cos^2 \gamma)\}}{\sin \beta} \qquad (4.14)$$

INTERPLANAR ANGLES

An advantage of Eqs. 4.9 to 4.12 is the facility they give to the calculation of interplanar angles. If we omit the $\lambda/2$ term and divide each residual x, y, z by the square root of the sum of their squares, we have a unit normal (i.e., one of unit length). For example, the $(12 \cdot 2)$ plane of lithium niobate has a unit normal

$$\begin{pmatrix} x \\ y \\ z \end{pmatrix} = \begin{pmatrix} \dfrac{1}{5.1494} \\[2mm] \dfrac{5}{5.1494 \times \sqrt{3}} \\[2mm] \dfrac{2}{13.8620} \end{pmatrix} \times \frac{1}{S}$$

where

$$S = \sqrt{\frac{1}{5.1494^2} + \frac{5^2}{5.1494^2 \times 3} + \frac{4}{13.8620^2}}$$

$$\begin{pmatrix} x \\ y \\ z \end{pmatrix} = \begin{pmatrix} 0.194197 \\ 0.560600 \\ 0.144279 \end{pmatrix} \times \frac{1}{\sqrt{0.194197^2 + 0.560600^2 + 0.144279^2}} = \begin{pmatrix} 0.31806 \\ 0.91815 \\ 0.23630 \end{pmatrix}$$

Similarly the (31·2) plane has a unit normal

$$\begin{pmatrix} x' \\ y' \\ z' \end{pmatrix} = \begin{pmatrix} 0.70937 \\ 0.68259 \\ 0.17568 \end{pmatrix}$$

The angle between two unit vectors $[x, y, z]$ and $[x', y', z']$ is Q, where

$$\cos Q = xx' + yy' + zz' \qquad (4.15)$$

This gives the scalar or dot product of two vectors. This product is not limited to unit vectors but is the triple product of the lengths times the cosine of the angle between them. Hence $\cos Q = 0.89386$ and the angle between the (12·2) plane and the (31·2) plane is 26°38′.

We can use a map of $(hk \cdot 0)$ planes to include all possible reflecting planes because every third level has the same x and y coordinates. Hence if we mark the zero level, the one level, and the two level with distinctive marks, any level l can be reached by adding $3n$ to the zero level, the one level, or the two level. However we must exclude odd values of l when h, k or $h + k = 0$. Figure 4.6 is such a map for lithium niobate. The limiting circle of such maps is really the largest cross section of a limiting sphere. The radius of the cross section for the level l appears in the scale at the bottom of Fig. 4.6. These radii are given by

$$R_l = \cos \sin^{-1}\left(\frac{l\lambda}{2c_0}\right) = \sqrt{1 - \left(\frac{l\lambda}{2c_0}\right)^2}$$

This scale quickly tells us how large l can be in $(hk \cdot l)$. We observe the distance of point $(hk \cdot 0)$ from the origin and lay out this distance from the arrow along the scale. The l cannot be larger than the next point to the right of the distance so layed out. For example, in $(15 \cdot l)$ we see that l cannot be larger than 5. Since (15·5) is permitted, this value could be used as an X-ray checkpoint, provided its Bragg angle is not too large. "Large" might mean greater than 80°. This reflection calculates out to be about 83.6°, too large for most goniometers.

We have extended the $(hk \cdot 0)$ plot for hexagonally indexed rhombohedral crystals to include all planes $(hk \cdot l)$. This was done by selecting three kinds of markings to identify "levels," that is, to show the remainder when the largest integer in $l/3$ is subtracted from $l/3$ for any given l value. Any point h, k on the plot is subject to the further restriction that if $h = 0$, $k = 0$, or $h + k = 0$, then l must be even. This is possible to do because the axis along which l is plotted is perpendicular to the plane of projection, the x, y plane. In attempting to extend the plot of $(0k \cdot l)$'s, to include all $(hk \cdot l)$'s, we are hindered because h is plotted along an axis that is not perpendicular to the plane of projection. As a

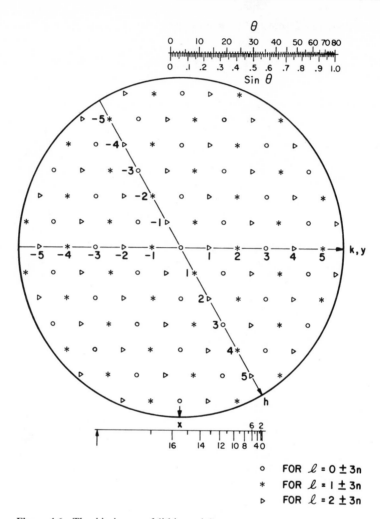

Figure 4.6   The $hk \cdot l$ map of lithium niobate.

result the y term for Cartesian coordinates contains both h and k, whereas the x term depends only on h, the z term only on l. We can remedy this by a change of variables. We introduce a new index j, defined as

$$j = \frac{h}{2} + k \qquad \text{and conversely} \qquad k = j - \frac{h}{2} \qquad (4.16)$$

We note that j can have "half" values: $j = 0, \pm\frac{1}{2}, \pm1, \pm1\frac{1}{2}$, and so on. In these terms the normal to the plane (hkl) Eq. 4.8 can be written

$$N(hjl) = \begin{pmatrix} h \\ \dfrac{2j}{\sqrt{3}} \\ \dfrac{la_0}{c_0} \end{pmatrix} \qquad (4.17)$$

Now j is plotted along the y axis and l is plotted along the z axis; h is thought of as plotted along x. Two symbols are needed to distinguish between integer j's and half-value j's. Such a plot for lithium niobate appears in Fig. 4.7. The rule that $-h+k+l = 3n$ now becomes $-3h/2 + j + l = 3n$ with the further restriction that if $h = 0$ or if $j = \pm h/2$,

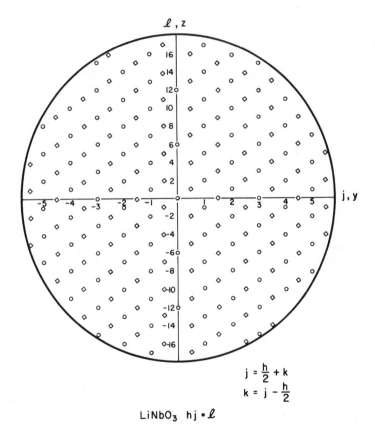

$$j = \frac{h}{2} + k$$
$$k = j - \frac{h}{2}$$

LiNbO$_3$ hj$\cdot l$

Figure 4.7 The $hj \cdot l$ plot for lithium niobate. The circles mark integer j's, the squares denote half-integer j's.

then l is even. As an example of the use of this chart we consider the planes near the (01·2) plane as possible orientation checkpoints. We note that if $h = 0$, the j notation is the same as the standard Miller-Bravais notation [i.e., the plane is also (01·2) on the j notation]. Two neighbors in the y, z plane of the (01·2) plane normal are the normals to the (03·9)$_j$ and the (04·5)$_j$ planes. These cannot be used because h is zero but l is not even. We try (05·4). This has a very large Bragg angle but satisfies the extinction conditions. Calculation gives the normal of (05·4) to be

$$N(05\cdot4) = \begin{pmatrix} 0 \\ 0.96844 \\ 0.24925 \end{pmatrix} \quad \text{while} \quad N(01\cdot2) = \begin{pmatrix} 0 \\ 0.84096 \\ 0.54109 \end{pmatrix}$$

This gives $\theta(05\cdot4) = 63°6'$ and by Eq. 4.15, $N(05\cdot4)$ is $18°20'$ away from $N(01\cdot2)$. On the other side of (01·2) (i.e., nearer the z axis), points with $h = 0$, l even, and near (01·2) are scarce. The nearest is (02·10). This has $\theta = 40°52'$ and is $25°53'$ away from (01·2).

Out of the y–z plane we find $(hj \cdot l) = (1, 7/2, 7)$ and $(\bar{1}, 7/2, 7)$. For these, $\theta = 47°15'$ and they are $11°45'$ on either side of (01·2). By Eq. 4.16 these are $(hk \cdot l) = (13\cdot7)$ and $(\bar{1}4\cdot7)$, respectively. Now $(13\cdot7) = (13\bar{4}7)$ cycles into $(\bar{4}137) = (\bar{4}1\cdot7)$. The order of h and k may be reversed if their signs are reversed—because of the mirror symmetry of this class. Hence (13·7) has the same properties as has $(\bar{1}4\cdot7)$.

FINDING ATOMIC PLANES NEAR ARBITRARY ORIENTATIONS

We are frequently obliged to map the normals of X-ray reflecting planes about an arbitrary direction—for example, planes near the normal of a desired crystal plate face. We assume that the lattice constants and space group, hence the extinction rules, are known, but charts of reflecting planes are not at hand.

We begin by calculating the direction cosines of the direction of interest. From Fig. 4.8 we see that the direction cosines of the direction D,

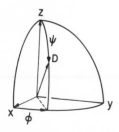

Figure 4.8 The direction defined by angles ϕ and ψ.

the direction of interest, are

$$D = \begin{pmatrix} \sin\psi\cos\phi \\ \sin\psi\sin\phi \\ \cos\psi \end{pmatrix} \tag{4.18}$$

We use the equations that give the normals to planes hkl and juggle values to find close matches to the expression for D.

For cubic crystals the problem is simple, since by Eq. 4.7 the normal to the plane hkl is $[h, k, l]$. To illustrate we choose the random values $\phi = 25°$, $\psi = 55°$, which gives $D = [0.74240, 0.34619, 0.57358]$. We divide the three terms by the smallest and get $[2.144m, m, 1.657m]$, where m is a factor of proportionality. If the smallest term is much smaller than the other two terms, we assume it to be near zero and divide the terms by the next smallest term. If m is taken as 1, we have the plane 211. If the crystal is body centered this reflects and can be used (if the crystal is primitive all hkl's are valid). For face-centered crystals 211 would not reflect, but 422 reflects and has the same orientation. The direction cosines of the normal to the 211 plane are

$$\frac{1}{\sqrt{6}}[2, 1, 1] \quad \text{or} \quad N(211) = [0.8165, 0.4082, 0.4082]$$

The angle between D and $N(211)$ is δ, where

$$\cos\delta = 0.7424 \times 0.8165 + 0.3462 \times 0.4082 + 0.5736 \times 0.4082 = 0.9816$$

which makes $\delta = 11°0'$. If we make $m = 2$ we obtain $hkl = 423$, which reflects only for primitive crystals. However we calculate $N(423) = [0.7428, 0.3714, 0.5571]$, which gives $\delta = 1.76°$. A 422 plane will reflect on most cubic crystals, again with $\delta = 11°0'$.

We complicate matters slightly by assuming the crystal to be rutile (tetragonal), $a_0 = 4.5929$ Å, $c_0 = 2.9591$ Å. Here by Eq. 4.9 we have

$$N(hkl) = \begin{pmatrix} h \\ k \\ \dfrac{la_0}{c_0} \end{pmatrix} = \begin{pmatrix} h \\ k \\ 1.5521l \end{pmatrix} \sim \begin{pmatrix} 0.7424m \\ 0.3462m \\ 0.5736m \end{pmatrix}$$

We divide the bottom terms by 1.5521 and obtain

$$\begin{pmatrix} h \\ k \\ l \end{pmatrix} \sim \begin{pmatrix} 0.7424m \\ 0.3462m \\ 0.3696m \end{pmatrix}$$

and dividing the last column by the middle term we get

$$\begin{pmatrix} 2.1444 \\ 1 \\ 1.0676 \end{pmatrix}$$

Obviously the 211 plane should give a fairly good approximation. Here

$$N(211) = \begin{pmatrix} 2 \\ 1 \\ 1.5521 \end{pmatrix} \quad \text{which normalizes to} \quad \begin{pmatrix} 0.7348 \\ 0.3674 \\ 0.5702 \end{pmatrix}$$

and departs from D by 1.32°. Copper $K\alpha$ X-rays reflect quite strongly from this plane with a Bragg angle of 27°10'.

An orthorhombic crystal is only slightly more involved. We consider iodic acid, for which $a_0 = 5.538$ Å, $b_0 = 5.898$ Å, and $c_0 = 7.733$ Å. Here by Eq. 4.10 we can write

$$\begin{pmatrix} \dfrac{h}{a_0} \\ \dfrac{k}{b_0} \\ \dfrac{l}{c_0} \end{pmatrix} = \begin{pmatrix} 0.18057h \\ 0.16955k \\ 0.12932l \end{pmatrix} \sim \begin{pmatrix} 0.7424m \\ 0.3462m \\ 0.5736m \end{pmatrix} \quad \text{so that} \quad \begin{pmatrix} h \\ k \\ l \end{pmatrix} \sim \begin{pmatrix} 4.111m \\ 2.042m \\ 4.435m \end{pmatrix}$$

Dividing the three last terms by the middle term, we have

$$\begin{pmatrix} h \\ k \\ l \end{pmatrix} \sim \begin{pmatrix} 2.01 \\ 1.00 \\ 2.17 \end{pmatrix}$$

which strongly suggests the (212) plane. The angle between D and the (212) plane is 2°3' by Eq. 4.15, and the Bragg angle is 21°28'. It reflects X-rays quite well.

A further complication enters with hexagonal crystals. Here

$$\left[h, \frac{h+2k}{\sqrt{3}}, \frac{la}{c} \right]$$

must approximate D; the y term contains both h and k, and the x and z terms each involve only one variable. We write

$$\begin{pmatrix} h \\ \dfrac{h+2k}{\sqrt{3}} \\ \dfrac{la}{c} \end{pmatrix} \sim \begin{pmatrix} D_1 \\ D_2 \\ D_3 \end{pmatrix} \quad \text{or} \quad \begin{pmatrix} h \\ h+2k \\ l \end{pmatrix} \sim \begin{pmatrix} D_1 \\ \sqrt{3}D_2 \\ \dfrac{D_3 c}{a} \end{pmatrix}$$

or

$$\begin{pmatrix} h \\ k \\ l \end{pmatrix} \sim \begin{pmatrix} D_1 \\ \dfrac{\sqrt{3}D_2 - D_1}{2} \\ \dfrac{D_3 c}{a} \end{pmatrix}$$

We illustrate with quartz. Here $a_0 = 4.903$, $c_0 = 5.393$; thus $c/a = 1.0999$ and $\lambda/2a_0 = 0.1571$.

$$\begin{pmatrix} h \\ k \\ l \end{pmatrix} \sim \begin{pmatrix} 0.7424 \\ -0.0712 \\ 0.6309 \end{pmatrix} \sim \begin{pmatrix} 1.176 \\ -0.113 \\ 1.000 \end{pmatrix}$$

Hence (10.1) seems to be a fair approximation.

$$N(10.1) = \begin{pmatrix} 1 \\ \dfrac{1}{\sqrt{3}} \\ 0.90914 \end{pmatrix} \dfrac{1}{1.4696} = \begin{pmatrix} 0.6804 \\ 0.3928 \\ 0.6186 \end{pmatrix}$$

$\sin \theta = 1.4696 \times 0.1571 = 0.2309$, $\theta = 13°21'$. The plane is $5°8'$ from D.

A rhombohedral crystal with hexagonal indexing is more difficult only because of greater extinction restrictions. An alternative way to analyze such a problem is through the use of j indices. We are reminded that $j = h/2 + k$ and can have half-values. The relation to be treated is

$$\begin{pmatrix} h \\ \dfrac{2j}{\sqrt{3}} \\ \dfrac{la}{c} \end{pmatrix} = \begin{pmatrix} D_1 \\ D_2 \\ D_3 \end{pmatrix}$$

thus

$$\begin{pmatrix} h \\ j \\ l \end{pmatrix} = \begin{pmatrix} D_1 \\ \dfrac{\sqrt{3}D_2}{2} \\ \dfrac{D_3 c}{a} \end{pmatrix}$$

For example, lithium niobate with $a_0 = 5.1494$ Å, $c_0 = 13.8620$ Å, $c/a = 2.692$, and $\lambda/2a_0 = 0.14958$, gives

$$\begin{pmatrix} h \\ j \\ l \end{pmatrix} = \begin{pmatrix} 0.7424 \\ 0.3074 \\ 1.5441 \end{pmatrix} \sim \begin{pmatrix} 1.000 \\ 0.404 \\ 2.089 \end{pmatrix}$$

which suggests $(hjl) = (1\frac{1}{2}2)$. That is, $(hk \cdot l) = 10 \cdot 2$. This fails the extinction rules but its third order $(30 \cdot 6)$ passes the test. The angle between $(30 \cdot 6)$ and D is $4°41'$ and $\theta = 38°2'$.

Having shown a method for finding planes nearly normal to an arbitrary direction, we wish to plot several such planes in the neighborhood to serve as checkpoints for a crystal surface orientation. To do this we add or subtract small integers to the $(hk \cdot l)$ set that was accepted as the nearest usable neighbor of D. The resulting $(hk \cdot l)$'s must not violate the extinction rules. This in general means that for rhombohedral crystals we can add ± 3 to $k + l$ and add any number to h if we also add it to $k + l$. We can shift numbers from k to l and from l to k. The presence of the mirror planes allows us to reverse the order of h and k if we reverse their signs. We will explore the neighborhood of $(30 \cdot 6)$ in lithium niobate. We add 1 to k and 2 to l to get $(31 \cdot 8)$, finding that

$$\cdot N(31 \cdot 8) = \begin{pmatrix} 0.5865 \\ 0.5644 \\ 0.5810 \end{pmatrix}$$

makes an angle of $15°25'$ with D and has a Bragg angle of $49°55'$. We shift 1 from l to k giving $(hkl) = (31 \cdot 5)$. This plane is $19°42'$ from D and has a Bragg angle of $43°0'$. Similar methods work readily on monoclinic and triclinic crystals.

IDENTIFYING STRANGE ATOMIC PLANES

If we rotate a crystal about an unknown plane normal and can identify two planes, the unknown can be found as their cross product. For example, a certain cubic crystal rotated about an unknown axis allows us to reflect X-rays from two different atomic planes. From the Bragg angles of these we identify them as (211) and (110), but the angle between them requires the (110) to be taken as $(1\bar{1}0)$ if the other identified plane is taken as (211). This is true because computation of the angle between (211) and (110) shows this angle to be $30°$, whereas the angle between (211) and $(1\bar{1}0)$ is computed to be $73.2°$. Hence if the observed angle is obviously nearer $73°$ than $30°$, we must use (211) and $(1\bar{1}0)$. Writing each digit twice in a row as before and striking out the end terms, we cross multiply:

$$
\begin{array}{c|cccc|c}
2 & 1 & 1 & 2 & 1 & 1 \\
& & \times & \times & \times & \\
1 & \bar{1} & 0 & 1 & \bar{1} & 0 \\
\hline
& & 1 & 1 & \bar{3} &
\end{array}
$$

Hence the plane must be $(11\bar{3})$. Since (113) and $(11\bar{3})$ do not reflect, we can identify the plane indirectly in this manner. Possibly $(22\bar{6})$ would reflect X-rays if the wavelength were short enough.

Non-cubic crystals call for slightly more complicated procedures, as previously indicated. We now give an example of plane identification from measurement of angles to any three noncoplanar identified planes.

An unknown plane has a normal $N_{(x)}$ that can be made into a unit vector through division by an unknown normalizing factor S. We write three equations of scalar products of $N_{(x)}$ and $N_{(i)}$ where $N_{(i)}$ is any one of three "normalized normals" of identified planes: $N_{(x)} \cdot N_{(i)} = S \cos G_i$. We then multiply both sides of one of these equations by a constant that makes the right-hand side of this equation equal to the right-hand side of another of the three equations. We then subtract the new equation from the one whose right-hand side was made to match, and subtract one from the other. This produces an equation $N_{(x)} \cdot N_{(i)} = 0$, which indicates that $N_{(x)}$ is perpendicular to $N_{(i)}$. We repeat with a different pair of the three identified equations getting a second equation, $N_{(x)} \cdot N_{(k)} = 0$, indicating that $N_{(x)}$ is perpendicular to $N_{(k)}$. Hence $N_{(x)}$ must lie along the cross product of $N_{(i)}$ and $N_{(k)}$. It is then a simple matter to determine possible values of h, k, and l using methods we have already illustrated.

We give as an example rutile, which is tetragonal with $a_0 = 4.5929$ Å, $c_0 = 2.9591$ Å. Hence

$$N_{(x)} = \begin{pmatrix} h \\ k \\ 1.55213l \end{pmatrix}$$

We assume that it has been observed that the unknown $N_{(x)}$ makes the following angles: $27.61°$ with the (310) plane, $11.53°$ with the (221) plane, and $43.53°$ with the (130) plane. Therefore we write

$$N_{(x)} \cdot \begin{pmatrix} 0.94868 \\ 0.31623 \\ 0 \end{pmatrix} = 0.88612S, \quad N_{(x)} \cdot \begin{pmatrix} 0.61990 \\ 0.61990 \\ 0.48108 \end{pmatrix} = 0.9798S,$$

and

$$N_{(x)} \cdot \begin{pmatrix} 0.31623 \\ 0.94868 \\ 0 \end{pmatrix} = 0.72501S$$

Multiplying the first of these equations by $0.97982/0.88612$ and subtracting it from the second, we find

$$N_{(x)} \cdot \begin{pmatrix} -0.42910 \\ 0.27023 \\ 0.48108 \end{pmatrix} = 0$$

Multiplying the third equation by 0.99982/0.72501 and subtracting it from the second equation, we have

$$N_{(x)} \cdot \begin{pmatrix} 0.19253 \\ -0.66220 \\ 0.48108 \end{pmatrix} = 0$$

The cross multiplication

$$\begin{vmatrix} -0.42910 \\ 0.19253 \end{vmatrix} \quad \begin{matrix} 0.27023 \\ -0.66220 \end{matrix} \diagdown \begin{matrix} 0.48108 \\ 0.48108 \end{matrix} \times \begin{matrix} -0.42910 \\ 0.19253 \end{matrix} \times \begin{matrix} 0.27023 \\ -0.66220 \end{matrix} \quad \begin{vmatrix} 0.48108 \\ 0.48108 \end{vmatrix}$$

gives 0.44857, 0.29905, 0.23212.

Hence $h:k:1.55213l = 0.44857:0.29905:0.23212$, and $h:k:l = 0.44857:0.29905:0.213212/1.55213$ or $h:k:l = 0.44857:0.29905:0.14955$.   We now plot these on a paper strip held against a log scale such as that of the d scale in Fig. 3.8. We then slide the paper along the scale, looking for a fit with integers, favoring smaller integers. With this example the fit is very good at 3, 2, 1. Hence the "unknown" is (321). It is advisable to indicate which plotted point is for h, which for k, and which for l, when plotting the numbers. Writing these letters over the plotted points avoids confusion.

Gnomonic Projection

To display the neighborhood of any particular direction, we can plot the points at which the normals of various permitted planes pierce a convenient plane of projection. This is a gnomonic projection. Projection on planes parallel to orthogonal axes is easy. We illustrate again with lithium niobate.

We divide the terms of the expression for hexagonal crystals by the y term, giving

$$\begin{pmatrix} x \\ 1 \\ z \end{pmatrix} = \begin{pmatrix} \dfrac{h\sqrt{3}}{h+2k} \\ 1 \\ \dfrac{(\sqrt{3}\, a_0/c_0)l}{h+2k} \end{pmatrix} \tag{4.18}$$

This gnomonic projection for lithium niobate is given as Fig. 4.9. Only reflections inside the limiting sphere are shown. One might ask why there are no reflections $0k \cdot l$ just above $01 \cdot 2$. A glance at Fig. 4.4 shows a clear field to $03 \cdot 12$. The angle scales, at the sides and bottom of the Figure indicate that $03 \cdot 12$ is more than $50°$ above the y axis.

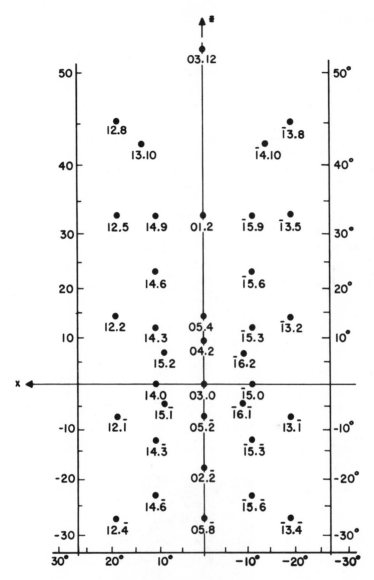

Figure 4.9 A gnomonic projection of lithium niobate. The plane of projection is 01.0.

Laue Photographs

Consider a small cylindrical beam of X-ray from an X-ray tube that puts out a broad range of wavelengths—for example, a tungsten target X-ray tube. If this beam falls on a crystal plate the beam will be reflected from many atomic planes in the crystal. A suitably placed photographic film can record the points at which the reflected beams pass through the film. Such a camera and a picture from it appear in Fig. 4.10. Note the fourfold symmetry in the picture. Laue photographs are used mostly for finding such symmetry directions. Much of the early work was done with very thin crystals, and the film was placed beyond the crystal to catch transmitted rays. However this transmission method requires crystals so thin that they are of little interest to those who intend to make such things as crystal oscillators from the plates so examined. Hence the back reflection method (Fig. 4.10) is the only practical method in crystal orientation work. In Fig. 4.11 an X-ray beam is striking a crystal plate normally. An atomic plane that is tipped at an angle ε with the surface will reflect the beam at an angle 2ε and will strike the film at a distance r from

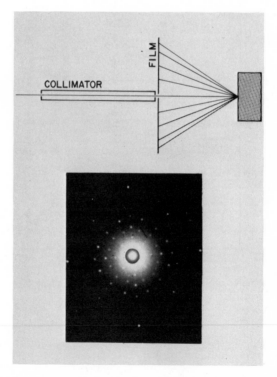

Figure 4.10 The Laue camera (top) and a Laue photograph.

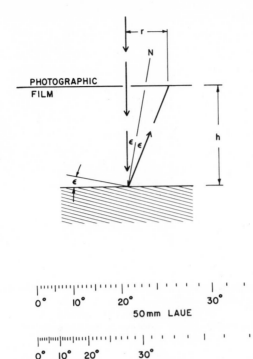

Figure 4.11   Analysis of a Laue photograph.

the point where the incident ray passes through the film. If h is the distance from film to crystal,

$$r = h \tan 2\varepsilon \qquad (4.19)$$

measurements on the film, together with the value of h, enable us to use Eq. 4.19 to calculate the angular amount by which a symmetry axis deviates from the direction of the incident X-ray beam. Calculations can be made separately for left-right deviations and for up-down deviations. At the bottom of Fig. 4.11 are two scales that measure these deviations directly, one for $h = 3$ cm and one for $h = 5$ cm, the two most commonly used distances. If the film is 10 cm long and $h = 3$ cm, then 2ε is less than 60°, which means that 2θ is greater than 120° or θ is greater than 60°. Let us suppose that the atomic plane is the $(00l)$ plane of rutile and that $\varepsilon = 5°$, giving $\theta = 85°$. Since the space group of rutile is $P4_2/mnm$, l must be even for planes of type $(00l)$. For rutile, $c_0 = 2.9591$ Å. Hence if $l = 2$, then by Bragg's law, λ must be $2d \sin \theta = 2(c_0/2) \sin 85° = 2.9478$ Å. If $l = 4$, however, λ would be 1.474 Å, and if $l = 6$ then λ would be 0.983 Å. If the

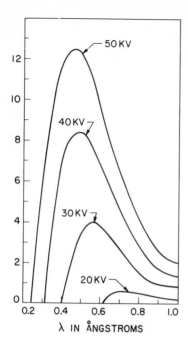

Figure 4.12 Intensity versus wavelength for a tungsten X-ray tube operated at several voltages.

wavelength spread of the X-ray generator is great enough, all these reflections would strike the same point on the film. However for λ = 2.95 Å air scatters X-rays drastically and badly fogs the film. The intensity of radiation from a tungsten target X-ray tube is represented as a function of wavelength in Fig. 4.12. It is seen that l = 12 would give maximum intensity for a tungsten tube operating on 40 kV.

The device in Fig. 4.13 allows us to go directly from the Laue X-ray equipment to a diamond saw. Two bars B, B straddle a track on the X-ray generator so that a crystal cemented to a rod R may be placed in the X-ray beam. The bars B, B are fastened to a disk D_1, graduated in degrees. A disk D_2 can be turned about the common axis of D_1 and D_2 and locked at any desired angle. Disk D_2 carries a plate P perpendicular to the plane of D_1 and D_2, and this plate, in turn, supports a disk D_3, graduated in degrees. Hence the crystal mounted on the rod can be adjusted to a desired orientation. Laue photographs taken of the crystal as mounted tell the proper angle settings for disks D_2 and D_3 to bring a symmetry axis parallel to the incident X-ray beam. From this information we can determine the angles at which these disks should be set for a given cut. When the disks D_2 and D_3 have been properly set, the device is placed in the saw and the cuts are made. With care the angles can be controlled to about a half degree of arc.

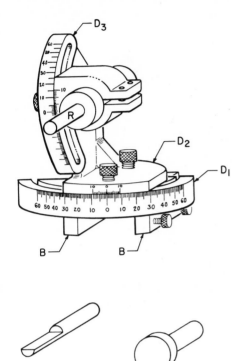

Figure 4.13   Device that can be transfer- . red to a diamond saw after X-ray orienting by means of Laue photographs.

Two of the several types of "rod" that facilitate the operation are illustrated in the bottom portion of Fig. 4.13.

Producing More Precisely Oriented Plates—The Barrel Holder

Over many years we have had much success with equipment (Fig. 4.14) in which a dovetail slide carries a "sled" that has a pair of vee notches into which a barrel holder is inserted. The barrel can be clamped down (clamps not shown) and the sled slid forward, causing the crystal to push a free sliding pin CP until a ring on the pin matches fiduciary marks (arrows). The front surface of the crystal is now centered on the axis of rotation of an arm that swings the slide about. An arc (indicated in figure), which is a curved dovetail, carries the detector and can be set to any 2θ angle between zero and 156°. Two adjusting screws AS1 and AS2 located on the end of the barrel allow the crystal to be tipped about two mutually perpendicular axes. In the position shown the screw AS1 allows adjustment about the vertical axis. On turning the barrel 90° in the vees, screw AS2 allows adjustment about the former horizontal axis, which has now become the vertical axis. When the adjustment is satisfactory, tightening

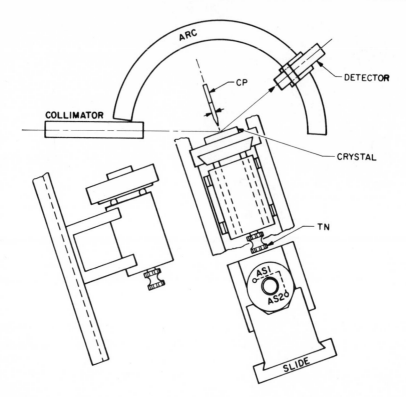

Figure 4.14 Practical crystal orientation goniometer. The barrel holder is adjusted in the goniometer, then transferred to a cutting machine.

the tension nut TN compresses an internal helical spring until a shoulder of the nut rests against the end of the barrel. The tilting plate on which the crystal is cemented is thus firmly locked. The holder now can be removed and transferred to a cutting machine. The "barrel holder" appears in more detail in Fig. 4.15.

Figure 4.16 presents a device that can be very useful for grinding a plane perpendicular to the barrel axis after X-ray adjustment. A sleeve is a smooth-sliding fit over a barrel such as the one of Fig. 4.15. The sleeve terminates in a square flange to which two equal-sided bars are fastened with screws. The barrel is thrust into the sleeve from below and the assembly "hand held" on a revolving lapping plate. The side bars maintain the perpendicularity of the barrel axis to the lap. Long slender bars can project from the ends of the space between the bars. For long slender bars a special "cap" fits over the tilting plate. This cap (Fig. 4.16, bottom)

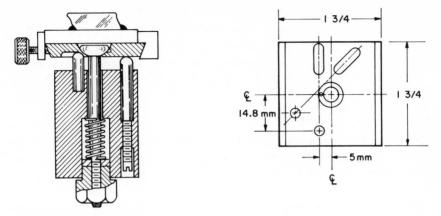

Figure 4.15 Details of the barrel holder.

supports the long crystal bar throughout its length. The steel side bars wear away much more slowly than do most crystals, hence they require little attention. If micrometer measurements show the bars to be no longer parallel and equal, they should be removed from the fixture and placed together on a surface grinder. Then a light cut is taken over the surfaces opposite to those that have the tapped holes.

This lapping device can be very useful for long slender bars. By

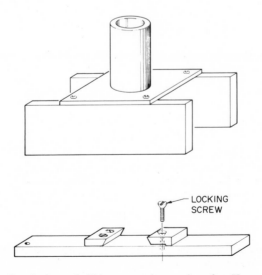

Figure 4.16 Grinding device to hold long crystals on a lap after X-ray orientation.

providing a second pit and slot on the barrel holder tilting plate (Fig. 4.15), the long crystal plates can be supported at 45° to the vertical, four ways, giving a complete single orientation. Such long crystals could not be oriented vertically because the ends would strike the goniometer. However with the 45° angle much longer crystals can be handled as shown in Fig. 4.17.

Stereographic Projection

The most commonly used projection in the field of crystallography is stereographic projection. All directions in three-dimensional space are here imagined as points on a sphere. Consider any given direction P, represented by the point p on a sphere Fig. 4.18. The point p is now joined to the south pole by a straight line that passes through the equatorial plane at p' and p' represents the direction P on the x, y plane. There are several advantages to this kind of projection. The upper hemisphere is represented inside a unit circle, that is, the cross section of the sphere (considered as having unit radius). The lower hemisphere can be also represented as within this circle if points are reflected in the x, y plane and marked with a different symbol (e.g., a small circle for directions in the upper hemisphere, small crosses for directions in the lower hemisphere that have been "reflected" into the upper hemisphere).

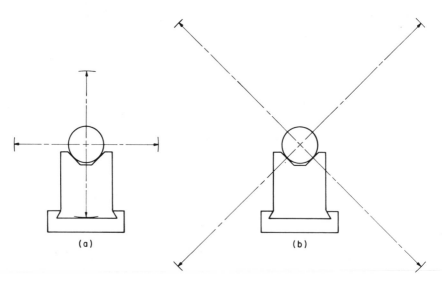

Figure 4.17 For extra-long crystals, a 45° mounting is possible with the barrel holder (Fig. 4.15) and the grinding jig (Fig. 4.16).

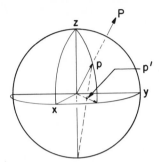

Figure 4.18 Derivation of the stereographic diagram to represent directions in space.

Gnomonic projection does not project a hemisphere into a finite plane. All circles on the sphere are projected as circles on the stereographic projection and therefore can be drawn with a compass. Angles are undistorted, and small areas are undistorted in stereographic projection.

For a point $[x, y, z]$ on a sphere of radius $\mathscr{R}$, $x^2 + y^2 + z^2 = \mathscr{R}^2$. This point is represented on a stereogram by

$$x' = \frac{x\mathscr{R}}{\mathscr{R} + z}$$

$$y' = \frac{y\mathscr{R}}{\mathscr{R} + z}$$

(4.20)

With this relation and Eqs. 4.7 through 4.12, stereographic projections of atomic plane normals can be drawn on ordinary graph paper for any class of crystal.

As an example let us plot the natural faces of the upper hemisphere of quartz. Quartz is trigonal, class 32, with $a_0 = 4.903$ Å, $c_0 = 5.393$ Å. The common faces are $m = 10\cdot0$, $r = 10\cdot1$, $z = 01\cdot1$, $x = 51\cdot1$, and hki may be cycled (i.e., rotated $\pm 120°$ about c). Here we have

$$N = \left[h, \frac{h + 2k}{\sqrt{3}}, \frac{la_0}{c_0} \right],$$

$$s = \sqrt{h^2 + \frac{(h + 2k)^2}{3} + \frac{l^2 a_0^2}{c_0^2}}$$

and

$$\left(\frac{a_0}{c_0} \right)^2 = 0.82654, \quad \text{also } \bar{N} = \frac{N}{S}$$

Face	$hk \cdot l$	N	S	$\bar{N}$	x', y'
m	10·0	$\left[1, \dfrac{1}{\sqrt{3}}, 0\right]$	$\dfrac{2}{\sqrt{3}}$	$[0.866, 0.500, 0]$	$[0.866, 0.500]$
r	10·1	$\left[1, \dfrac{1}{\sqrt{3}}, 0.9091\right]$	1.470	$[0.680, 0.393, 0.619]$	$[0.420, 0.243]$
z	01·1	$\left[0, \dfrac{2}{\sqrt{3}}, 0.9091\right]$	1.470	$[0, 0.786, 0.619]$	$[0, 0.485]$
s	11·1	$\left[1, \dfrac{3}{\sqrt{3}}, 0.9091\right]$	2.197	$[0.455, 0.788, 0.414]$	$[0.322, 0.557]$
x	51·1	$\left[5, \dfrac{7}{\sqrt{3}}, 0.9091\right]$	6.493	$[0.770, 0.622, 0.140]$	$[0.675, 0.546]$

These values and their cycled equivalents are plotted in Fig. 4.19 along with a clinographic drawing of right quartz. Since faces m, x, s, z, r, s, m all lie on a circle made with a compass, we see that all these faces are parallel to one line, the "zone axis"; thus the faces m, x, s, z, r, s, m are said to belong to one zone. Since the crystal is trigonal, there are three such arcs.

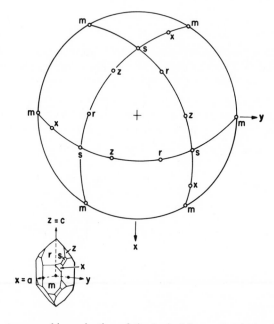

Figure 4.19 A stereographic projection of the typical faces of right-hand quartz.

The zone axis is a vector parallel to all faces in the zone. We can find this vector by taking a "vector product" of any two face normals. This is done, as before, by writing each one twice, one set under the other, then crossing out the end terms and cross multiplying as in the example:

$$P_1P_2P_3P_1P_2P_3$$
$$\times \times \times$$
$$P'_1P'_2P'_3P'_1P'_2P'_3$$
$$P_2P'_3 - P'_2P_3, \; P_3P'_1 - P'_3P_1, \; P_1P'_2 - P'_1P_2$$

If the P's are Miller indices or Miller-Bravais indices, the zone axis will be expressed in terms of crystallographic axes. If the P's are xyz coordinates of plane normals, the zone axis will be expressed as xyz coordinates.

As an example the zone defined by $s = 11\cdot1$ and $x = 51\cdot1$ is

$$1\;1\;5\;1\;1$$
$$\times$$
$$1\;1\;1\;1\;1$$
$$1\text{-}1, 1\text{-}5, 5\text{-}1 = 0\bar{4}\cdot4 \quad \text{or} \quad 0\bar{1}\cdot1$$

that is, a vector defined by one backward step along the a_2 axis and one forward step along the c axis of the crystal.

Let us consider a stereogram of which the radius of the unit circle is $\mathcal{R}$. On this stereogram a point p' is at a distance "a" from the y axis and at a distance "b" from the x axis. Hence $x' = a/\mathcal{R}$, $y' = b/\mathcal{R}$. The point p' is at a distance f from the origin and $f^2 = x'^2 + y'^2$. From these measurements we can find the elements of P in three-dimensional space. If P' makes an angle σ with the z axis, then

$$\tan\frac{\sigma}{2} = \frac{f}{\mathcal{R}} \tag{4.21}$$

$$z = \cos\sigma \tag{4.22}$$

From these we calculate

$$x = x'(1+z), \qquad y = y'(1+z) \tag{4.23}$$

The importance of this step is in measuring the angle between any two points on the stereogram. Consider the two points $r = 10\cdot1$ and $s = 11\cdot1$ for which $[x', y']$ were [0.420, 0.243] and [0.322, 0.557], respectively. For the first $f = 0.487$, for the second $f = 0.643$. Hence for the first $\sigma = 51.9°$ for the second $\sigma = 65.5°$. For the first $z = 0.617$, for the second 0.415. Finally, the first is $[x, y, z] = [0.421 \times 1.617, 0.243 \times 1.617, 0.617]$. For the second $[x, y, z] = [0.322 \times 1.415, 0.557 \times 1.423, 0.415]$. This gives [0.681,

0.393, 0.617] for the first, [0.456, 0.790, 0.415] for the second. The cosine of the angle between r and s is consequently $0.681 \times 0.456 + 0.393 \times 0.790 + 0.617 \times 0.415 = 0.877$ and the angle is 28.6°.

Great Circles and Small Circles of a Sphere

A plane that passes through the center of a sphere intersects the spherical surface in a circle called a "Great Circle." If a plane cuts through a spherical surface but does not pass through the center, it makes a smaller circle called a "Small Circle."

A great circle and a small circle are given in stereographic projection in Fig. 4.20. Also shown is the radius with which each is to be drawn and the position of the centers of the arcs. For the great circle at angle ψ

$$R = \frac{r}{\sin \psi} \tag{4.24}$$

$$D = r\left(\frac{1}{\sin \psi} - \frac{\tan \psi}{2}\right) \tag{4.25}$$

For a small circle at angle ρ

$$R' = r \cot \rho \tag{4.26}$$

$$D' = r\left(\cot \rho + \tan \frac{\rho}{2}\right) \tag{4.27}$$

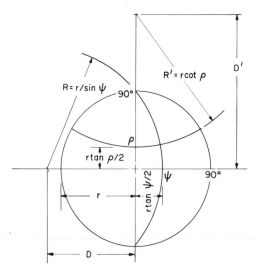

Figure 4.20 Stereograph of great circle and small circle.

It is often desirable to project onto some plane other than the x, y plane (001). This is easily done by cycling x, y, z. On maps of the Western hemisphere of the earth, the [00$\bar{1}$] point called the South Pole is replaced by the point at which the international date line crosses the equator. The great circles are then meridians of longitude, the small circles are parallels of latitude.

Stereographic protractors are available (see, e.g., Fig. 4.21) for measuring the angle between two plane normals—if the unit circle of the protractor has the same radius as has the unit circle of the stereogram.

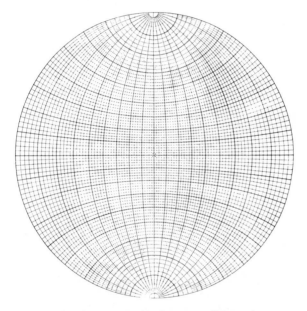

Figure 4.21 A stereographic "protractor" (or net).

We place the transparent protractor over the stereogram with the centers superimposed and turn the protractor about until the two points lie on the same meridian, as in Fig. 4.22. Reading the two parallels, we then take the difference, which is the angle between the two planes.

Greninger Nets

Just as a stereographic protractor can be used to read the latitude and meridian of any point on a stereographic projection—if they have the same size unit circles—a Greninger net can be placed on a Laue photograph to permit reading the latitude and meridian for any spot—if

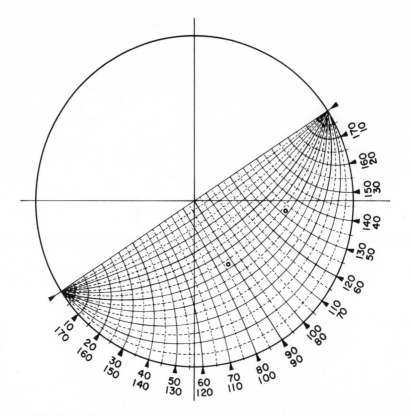

Figure 4.22 Measuring the angle between two directions on a stereographic projection.

they are made for the same crystal to film distance. In Fig. 4.23*a* we see an atomic plane unit normal *N* with latitude *L* and meridian *M*. X-rays are considered as traveling along the *x* axis toward the origin and being reflected by the plane of normal *N* along the direction *x'* (Fig. 4.23*b*). The *x, y, z* coordinates of *N* are

$$N = \begin{pmatrix} \cos L \cos M \\ \cos L \sin M \\ \sin L \end{pmatrix} \tag{4.28}$$

The sine of the Bragg angle is

$$\sin \theta = \cos L \cos M \tag{4.29}$$

Figure 4.23*b* indicates that the vector *x* plus the vector *x'* is

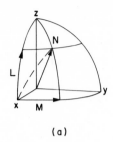

(a)

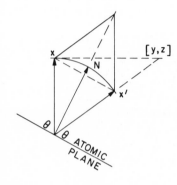

(b)   **Figure 4.23** Analysis of the Greninger net.

$2 \cos L \cos M$ times the vector N. Hence

$$x' = \begin{pmatrix} 2 \cos^2 L \cos^2 M - 1 \\ 2 \cos^2 L \sin M \cos M \\ 2 \sin L \cos L \cos M \end{pmatrix} \qquad (4.30)$$

If the film is placed in the (100) plane, the Laue spot will appear with the following coordinates:

$$\begin{pmatrix} y \\ z \end{pmatrix} = \frac{\cos L \cos M}{(\cos L \cos M)^2 - 1/2} \begin{pmatrix} \cos L \sin M \\ \sin L \end{pmatrix} \qquad (4.31)$$

If the film is at a distance H from the crystal, the y and z coordinates must be multiplied by H. The Greninger net (Fig. 4.24) is made by plotting on graph paper values of L and M as functions of y and z of Eq. 4.31.

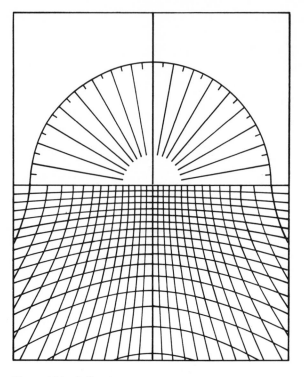

Figure 4.24 A Greninger net.

Complete Orientation

Thus far we have adjusted the crystal for only one atomic plane, then cut or lapped a face. It is then possible to remount the crystal, adjust it for a second atomic plane, and cut a second face parallel to the second atomic plane. The two faces should be marked with the indices of the two atomic planes. From these two marked faces, any orientation can be produced by mechanical devices without further use of X-rays. Adjustment for only one plane does not give a complete orientation.

Sometimes it is desirable to know the complete orientation within a couple of degrees before cutting is started. One simple subterfuge is to orient one atomic plane and mark it before shifting to a position suitable for finding the second plane. This can often be done with a felt pen, either marking the trace of the plane or indicating the plane normal, preferably the latter. For irregularly shaped crystals it is more accurate to attach a toothpick to the crystal with soft wax. This can easily be made perpendicular to the atomic plane within 2°. The crystal is then carefully repositioned so that the sweep arm rotates the crystal about an axis parallel to the toothpick. This second plane is then searched out and similarly marked with a toothpick of a different color. Large carpet tacks also serve, the head giving a larger surface to stick to the wax, but color coding carpet tacks is not as easy as putting ink spots on wooden toothpicks.

Sometimes it is desirable to adjust a crystal to a holding fixture with two atomic planes simultaneously in adjustment. A simple example is the complete orientation of small ferrite spheres to be placed in a magnetic field wave guide in a definite orientation. The device in Fig. 4.25 can be used if the two atomic planes are mutually perpendicular. The sphere is "mounted"; that is, it is stuck to the sharp end of a pointed 1/8 in. diameter metal rod or "pin." The pin is thrust into the bushing, and the amount left projecting ensures that the sphere is centered on the barrel axis, as in Fig. 4.25. The barrel is placed in the goniometer vees with the sphere in the X-ray beam, and the detector is set for some atomic plane of convenience. The goniometer arm is then slowly swung around, searching for a reflection. If none is found, the pin bushing is turned a few degrees

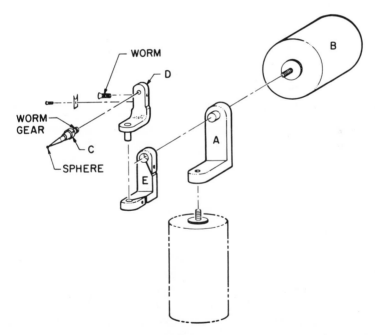

Figure 4.25 Device for completely orienting a single crystal sphere: A is the angle piece, B is the barrel, C is the bushing, D is the small link, and E is the large link.

with a screw driver and another sweep is made. This procedure is continued until a reflection is found. When the reflection is found, the bushing and the small link are adjusted until the barrel can be rotated 360° in the vees without losing the reflection; the arm reading will then be the Bragg angle θ of the "plane of convenience." When this is achieved the angle piece is removed from the barrel and repositioned, using the tapped hole at the other end of the angle piece A. Now a search can be made about the first atomic plane normal as an axis. When the second atomic plane is found, the large link is adjusted to make the arm read the second atomic plane Bragg angle. The barrel can then be rotated 360° in the vees without losing the reflection. After this double adjustment is perfected, the sphere is transferred to another pin, using a cement that is insoluble in a solvent that can dissolve the cement used for the first mounting. Fixtures for the first mounting of the sphere on the pin and for facilitating the transference of the sphere to the second pin appear in Figs. 4.26 and 4.27, respectively. The insert of Fig. 4.27 shows the purpose of the equipment. A tiny sphere a half-millimeter in diameter is mounted on a pin with a 111 axis along the pin axis and a 110 plane parallel to a flat on the pin. It is now ready to insert in experimental apparatus at known orientations.

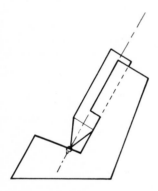

Figure 4.26 Fixture for mounting a sphere on the tapered end of a $\frac{1}{8}$ in. pin.

For larger but irregularly shaped crystals a similar device (Fig. 4.28) is used. Since the crystal is now not a sphere, we must provide for frequent recentering during the sweeps. The sweeping is most productive if a rounded corner is centered on the axis so that centering persists during a rather large angular sweep. The crystal in Fig. 4.29 could well be centered at any one of the several crosses shown. A very sharp edge should be slightly rounded to facilitate reflection.

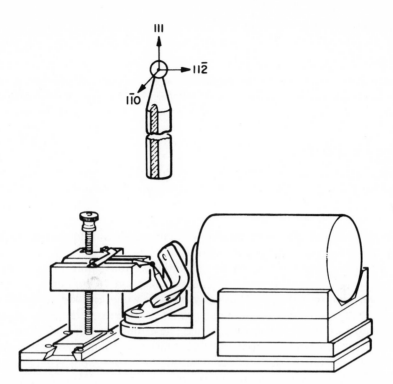

Figure 4.27 Fixture for transferring the sphere to a second pin. A known atomic plane of the sphere is perpendicular to the pin axis, and a known plane is parallel to the flat on the side of the pin.

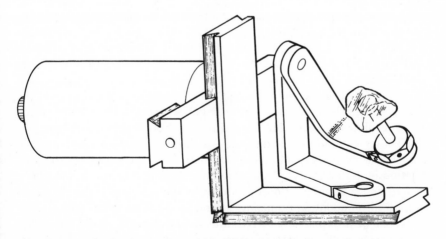

Figure 4.28 Device for the complete orientation of a larger but irregularly shaped crystal.

Figure 4.29 Rounded edges are centered to increase the sweep range when searching for reflections.

Precision of Orientation

The Laue photograph method of orientation is accurate to about $\frac{1}{2}°$ as generally used, $\frac{1}{4}°$ if especial care is exercised. The goniometer method is more precise, and we now examine its practical limits of precision. Assuming a perfect crystal, the limits of precision are set primarily by the wedge angle of the rays passed by the collimator and also by the wavelength spread of the radiation used.

In any discussion of X-ray wavelengths we must note that the early determinations, made from chemical considerations, were not as accurate as later determinations. The early values were calculated from the number of sodium chloride atoms per unit volume using the best value of Avogadro's number. They were called angstroms but were later found to be in error in a ratio of about 1.00202. It has been found to be convenient to continue using the data collected before the error ratio was established, but to call the unit not an angstrom but a kXU (kilo X-unit). The correction was made when it was found possible to diffract X-rays from a very fine ruled grating that could also be used for light diffraction. This tied the units to the cgs system. Often the symbols Å and kXU are used indiscriminantly. When lattice constants are stated, no harm is done if the wavelength of the X-rays used to establish them is given. In these terms, copper radiation has three main wavelengths, $Cu\,K\alpha_1$ is of length 1.537395 kXU (= 1.540562 Å) with a half-width at half-peak intensity of 0.00029 kXU. However $K\alpha_1$ has a near neighbor, $K\alpha_2$ of wavelength 1.541232 kXU with half-width at half-maximum intensity of 0.000384 kXU. The second wavelength has about half the intensity of the $K\alpha_1$ radiation. There is also a $K\beta$ radiation at about 1.38935 kXU, which can be resolved into $K\beta_1$ and $K\beta_2$ with $K\beta_1$ about 100 times as strong as $K\beta_2$. These are a little wider than the $K\alpha$ lines but can be filtered out. The filter should be nickel foil, about 0.02 mm thick. This reduces the $K\alpha_1$ radiation by a factor of 2.15 but it reduces $K\beta$ by a factor of 133. Since the $K\alpha_1$ line is about 5.5 times as strong as the $K\beta_1$ line in the unfiltered radiation, it is about 340 times as strong in the filtered radiation.

Because of the closeness of $K\alpha_1$ and $K\alpha_2$, two peaks of the rate meter may be seen as the goniometer arm passes through a Bragg reflection. Since $\lambda = 2d \sin \theta$, we see that a small change in wavelength $\Delta\lambda$ changes the Bragg angle by $\Delta\theta$ where

$$\Delta\theta = \frac{\Delta\lambda}{2d\cos\theta} = \frac{\Delta\lambda}{\lambda}\tan\theta$$

For $\Delta\lambda = (1.541232 - 1.537395) = 0.003837\,\text{kXU}$, we see that here

$$\frac{\Delta\lambda}{\lambda} \sim 0.0025$$

Hence at $\theta = 20°$ the peaks for $\text{Cu}\,K\alpha_1$ and $\text{Cu}\,K\alpha_2$ are separated by $0.0025\tan 20°$ radian (i.e., 0.000908 radian, which is 3.1 minutes of arc). If the collimator wedge angle exceeds this angle, which it generally does, $\text{Cu}\,K\alpha_1$ and $\text{Cu}\,K\alpha_2$ are not resolved. In this case we might do better to weight the $K\alpha_1$ and $K\alpha_2$ in the ratio of 2:1 and use the weighted mean as the effective wavelength (i.e., $\lambda = 1.5387\,\text{kXU}$). More accurately, the weighting should be 2.09:1. This gives $\lambda = 1.53863\,\text{kXU}$. If the Bragg angle is 45°, then $\Delta\theta = 0.0025$ radian $= 8.6$ minutes of arc.

The width of the $K\alpha_1$ radiation at half maximum, divided by the wavelength, is

$$\frac{\Delta\lambda}{\lambda} = 0.000377$$

At $\theta = 20°$ this gives $\Delta\theta = 0.001035$ radian $= 3.6$ minutes of arc. The center of this can easily be found to about one-tenth of this amount, or about 22 seconds of arc. This gives some idea of the precision that can be obtained with the goniometer method.

Collimator Wedge Angle

A collimator formed by two slits of width w separated by a distance L has a wedge angle $\varepsilon = 2w/L$ radians. Hence for slits 100 mm apart to give $\varepsilon = 3$ minutes, we need w about 0.044 mm. Such fine slits give weak signals, and a rate meter can become jumpy due to statistical variations in the arrival rate of quanta into the detector. This disadvantage can be overcome at the cost of time by using a system that counts the quanta received in a fixed time, say 10 seconds. A plot of quanta per unit of time versus angular setting allows one to evaluate θ to a few seconds of arc—assuming that the goniometer angle scale is precise enough to work in that range.

To work to such small angular tolerances requires temperature control. For example, the coefficient of expansion of crystal quartz is about 14.6×10^{-6} per degree centigrade perpendicular to the optic axis. Differentiating $\lambda = 2d\sin\theta$ with λ constant, α and θ varying we have

$$\frac{\Delta d}{d} = \frac{-\Delta\theta}{\tan\theta}$$

Hence

$$\Delta\theta = \frac{-\Delta d}{d}\tan\theta$$

But

$$\frac{\Delta d}{d} = 14.6 \times 10^{-6} \times \Delta t$$

where Δt is the temperature change. The 11·0 reflection of quartz at room temperature is about 18°17′, the 22·0 reflection comes at 38°53′, and the 33·0 reflection at 70°15′. Hence the change in Bragg angle due to a temperature change of 10°C is 10 seconds of arc for (11·0), 37 seconds for (22·0), and 1 minute 25 seconds for (33·0). The coefficient of expansion along the optic axis of quartz is 7.8×10^{-6} per degree centigrade. Calling the coefficient along the axis a_3 and the coefficient normal to the optic axis a_1, we find that at an angle φ from the optic axis the coefficient is $a_\varphi = a_3 + (a_1 - a_3)\sin^2\varphi$.

Refraction

Another source of possible error is refraction. The X-rays bend as they enter the crystal. The refractive index is of the order of $(1 - 10^{-5})$. The wavelength inside the crystal is consequently longer than it is outside, and Bragg's law applies inside, not outside. A rough relation is that the refractive index is $1 - 4.48 \times 10^{-6}\,N_0\lambda_0^2$ where λ_0 is the wavelength in kXU outside and N_0 is the number of orbital electrons per cubic kXU in the crystal. The value of N_0 is found by summing the atomic number Z of the elements per unit cell and dividing by the cell volume in cubic kXU. For example, gallium arsenide has four gallium and four arsenic atoms in a cell 5.64 kXU on a side. Hence the refractive index is approximately $1 - 4.48 \times 10^{-6} \times 4(31 + 33) \times 1.54^2/5.64^3 \sim 1 - 15 \times 10^{-6}$. As a result the Bragg angle is reduced by an amount $\varepsilon = \delta/\tan g$, where g is the grazing angle, which might not be the Bragg angle. The Bragg angle for gallium arsenide (111) reflection is thereby reduced by 12.7 seconds of arc from about 13°41, the uncorrected value. It is seen that the correction is very small unless the angle g is small and the crystal is dense. The grazing angle can be very small for surfaces tipped far from the atomic reflecting plane. Let us consider bismuth germanium oxide, $Bi_{12}GeO_{20}$. Here $\delta = 24.2 \times 10^{-6}$ for copper Kα radiation. The (110) reflection occurs for $\theta = 6°10′$. The angle correction for planes parallel to the surface is about 50 seconds of arc. If the surface is 5° from the (110) plane, however, the grazing angle is $g = 1°10′$ at its minimum. The correction is then 4.1 minutes of arc. As the crystal is rotated about the atomic normal, this correction falls as g increases.

Tilt Error

A further source of error is "out of plane tilt." If the normal to the plane of reflection is not perpendicular to the axis of rotation (and the X-ray beam) but fails this by a small angle φ, there is an apparent Bragg angle θ' where $\sin \theta = \cos \varphi \sin \theta'$. A simple approximation shows the error in θ to be

$$\Delta\theta = \frac{\varphi^2}{2} \tan \theta \text{ radians}$$

where φ is assumed to be measured in radians. For example, in a crystal tilted by 5° "out of plane," a θ of 20° would appear as 20°4.8′. In round numbers, a tilt of t degrees makes the Bragg angle look larger than its true value, larger by $(t^2 \tan \theta / 115)°$.

Another result of this effect is a slit height error. The slits may be narrow but long. Contrary to our assumptions, rays that enter the top of the first slit but leave near the bottom of the second slit do not strike the crystal at the angle we have assumed. Slits 1 cm high and 10 cm apart have a limiting φ of arctan 1/10 = 5°43′ and should require a correction of 6.24 minutes of arc. This will be averaged in with rays making less extreme passages. This averaging gives an error in θ of

$$\epsilon = \frac{\varphi^2}{6} \tan \theta$$

Hence the last-named collimator would falsify θ by 2.1 minutes of arc. If the slits are not parallel to the radiating area of the X-ray tube target but normal to it, this effect is much reduced.

There are two other corrections that might be made, the Lorentz and the polarization correction, but both are negligibly small.

We reemphasize that all the foregoing assumes a perfect crystal. Crystals that have been sawed, lapped, or polished may have a disturbed surface that broadens the reflections. Etching generally takes care of this. Removal of 0.05 mm is often enough.

Crystal Perfection

Although crystals may have a very high degree of perfection, most show imperfections if subjected to sufficiently exacting tests.

All crystals have vacancies—lattice sites where an atom should be, but is missing. All crystals have impurities—sites where a foreign atom substitutes for the one that should be there. Foreign atoms are often jammed in between lattice sites, where they crowd their neighbors but do not replace one of them; these are interstitial impurities. Crystals have stacking faults and dislocations.

The phenomenon of "twinning"—two crystals grown together—is often obvious to the naked eye. As the crystals were growing, a disturbance in growth conditions caused the change in stacking. The new part may be a mirror image of the old. The surface where the two parts join is called the composition face. It may or may not be a mirror plane that relates the two parts by reflection symmetry. The composition face may be far from planar, but it is made up of atoms that belong equally to both parts of the crystal. Sometimes it is more logical to think of the two parts as related by rotation, not reflection. This is true of crystals that have a screw axis. If both parts have right-handed screw axes, for example, they could not be related by reflection because the mirror image of a right-handed screw is a left-handed screw.

Etching

Several kinds of twinning can be studied in quartz. Below 575°C quartz has trigonal symmetry, above 575° it has hexagonal symmetry. As quartz cools down through the transition temperature, outlying portions pass through the transition before the main body does. These outlying portions may not all lose the same three axes, and as the transition sweeps through the body, the portions meet but disagree about the positive end of the "a" axes. This can be seen by etching a plate and viewing the surface, preferably by *one* not-too-near-light source. Etch pits have different slopes in the two kinds of regions and this causes differences in reflection. Sharp boundaries separating the two kinds are visible. Sand blasting achieves the same effect for some crystals, but more quickly, but it works only on crystals that have cleavage planes.

Another kind of twinning often found in quartz is "right–left twinning." Here regions of one kind penetrate regions of the other kind. This condition is generally obvious in (00·1) sections placed between crossed polarizers.

Laue Photographs

Laue photographs can be made of various parts of the crystal. If a region covered by the X-ray beam has slightly misoriented material, the spots on the photograph will be multiple, as in Fig. 4.30. If different parts of the crystal exhibit quite different patterns, the crystal may be twinned or it may have lineage. Many crystals "pulled from a melt" have misoriented material near the surface where thermal shock may be great. This is "lineage." If the Laue camera incident beam straddles a twin boundary, two sets of Laue patterns are superimposed.

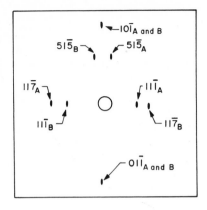

Figure 4.30 Tracing from a Laue photograph of a composite crystal. Parts A and B both record on the film.

Goniometer Scanning

Another method for checking crystal perfection is by observing the arm readings on the goniometer as the crystal is slid along a straight track. Some apparently perfect crystals have been found to have atomic planes that deviate more than 5° in 15 cm of length. However deviations of a fraction of a degree are easily observed on the goniometer. To facilitate this investigation, researchers have used a scanning head, which fits on the standard barrel holder and allows for translation in two mutually perpendicular directions; scales enable the translations to be measured. It

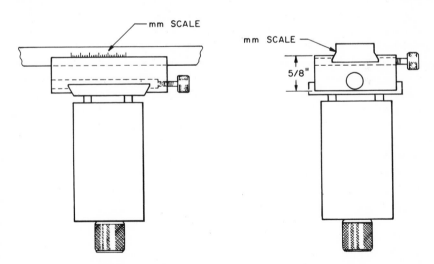

Figure 4.31 Scanning head permits quick scanning of a large crystal, to detect "bends" in atomic planes.

is best to replace the slit collimator with a pinhole collimator. The scanning head is illustrated in Fig. 4.31. To be thorough, the crystal should be rescanned after turning it 90° in the vees, because small bends about the axis of rotation (AR) of the goniometer are more readily detected than small bends about an axis at right angles to AR.

The Double Crystal Goniometer

For a more exacting test of crystal perfection we can use a double crystal goniometer (Fig. 4.32). Rays from the X-ray tube target T strike the first crystal A, reflect onto the second crystal B, and from B reflect into the detector D. When D is near a maximum counting rate, a photographic film is placed at F. Development often reveals that the reflecting atomic plane is not planar. It should be emphasized that crystals A and B should have nearly the same Bragg angle.

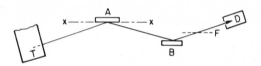

Figure 4.32 Double crystal goniometer arrangement used to detect "bends" in atomic planes.

If the reflecting planes of crystal A are tipped relative to the crystal surface as in Fig. 4.33, a wider photograph is obtained. Crystal A should be adjustable about the axis $X \ldots X$ (Fig. 4.32), as well as about an axis $Y \ldots Y$ perpendicular to the paper. With crystal B removed, crystal A should be adjusted about $Y \ldots Y$ until the detector shows the strongest reflection. Crystal B is then repositioned and adjusted for strongest reflection. Now it should be possible to rotate crystal B about an axis perpendicular to the reflecting planes without losing the signal. When this is accomplished, slight adjustment of crystal A about axis $X \ldots X$ should enhance the final reflection from B. Such reflections may be only a few seconds wide at half-maximum.

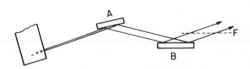

Figure 4.33 Broadening the coverage of crystal B by using tilted planes in crystal A.

Although reflections from crystal B can be obtained with the configuration of Fig. 4.34, this is not an effective way to examine crystal perfection. We call this the antiparallel configuration and the configuration of Fig. 4.32 the parallel configuration.

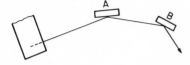

Figure 4.34 The antiparallel arrangement, not good for crystal perfection studies.

Coming from crystal A is a reflection whose angular spread is shown exaggeratedly in Fig. 4.35 and again in Fig. 4.36. This angular spread is partly due to the spread in wavelengths in the radiation—"copper $K\alpha_1$" has a half-width at half-maximum of about 190 parts per million. The angular spread is also due partly to crystal imperfection. In crystal imperfection we include not only deviations of atomic planes from strict

Figure 4.35 The angular spread in reflection of $K\alpha$ X-rays from a crystal surface.

planarity but also the deviation of atoms from their rest positions due to thermal vibrations of atoms. Crystal B has a similar angular spread. In rotating crystal B through a reflection with the configuration of Fig. 4.35 we are, in a sense, sliding the angular spread function of crystal B over the spread function of crystal A. In the antiparallel configuration the spread

Figure 4.36 The angular spread of Fig. 4.35 plotted linearly.

function of crystal B is reversed, and the curves cannot match except over very narrow wavelength ranges. This can be understood by examining Fig. 4.37, where it is clear that a maximum will be found where the α_1 peaks match. However the shorter wavelength side of crystal B will be over the longer wavelength side of crystal A, and the reflection will be

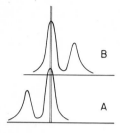

Figure 4.37 Sliding one spread function over another as in the double crystal goniometer.

diminished. With the parallel configuration, all parts match simultaneously. This shows why the Bragg angles of the two crystals should match rather well. If the Bragg angles do not match, the two curves cannot match everywhere at once.

If we plot counting rate against angular positions of crystal B, we get a "rocking curve." The width at half-maximum of such a curve is inversely proportional to sin 2θ. For some crystals this width is only a few seconds for θ near 45°, the optimum angle for sharpness. Crystals with wide rocking curves can be photographed by placing a film at F (Fig. 4.32). Pictures taken at different parts of the rocking curve reveal that different parts of the crystal reflect at different angular settings.

Figure 4.38 presents a sample cut from a germanium crystal "pulled from a melt." X-rays were reflected from the 111 face, and the rocking curve was taken for rotation about a [110] direction. Its width at half-maximum was about 20 seconds of arc. A picture taken at the peak appears in insert a. Halfway down the curve a second picture was taken (insert b). The crystal was then turned 90°, giving the rotation axis a [211] direction. The rocking curve, about 25 seconds wide at half-maximum, is shown in Fig. 4.39. Insert a is a picture taken at the rocking curve peak;

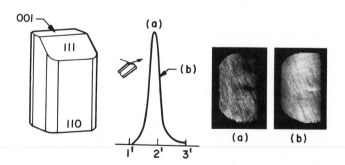

Figure 4.38 Perfection study of a germanium crystal, two views taken a fraction of a minute of arc apart.

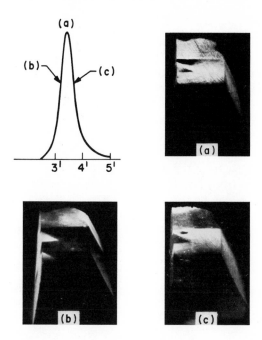

Figure 4.39 Further study of the germanium crystal of Fig. 4.38, turned 90°. Three views are taken, a fraction of a minute of arc apart.

pictures taken halfway down on each side appear as inserts *b* and *c*. While the crystal was being pulled it lost contact with the molten material because the pulling proceeded too rapidly. Within a few seconds the solid part was lowered into the melt, and growing continued. The section photographed in Figs. 4.38 and 4.39 cuts through this interrupted region. In Fig. 4.38 we see a "knot" about where it would be expected. We also see grinding scratches that did not etch clean. However examination of this area at 90° from the first view reveals that there is much more to be seen.

The left-hand side of the three pictures of Fig. 4.39 is caused by reflection from 111 planes projecting onto the 001 face. The right-hand side of the three pictures is caused by reflection from 111 projecting onto the 110 face. The central area is caused by reflection from 111 planes as seen on the 111 face. The geometry of the projection explains the large vertical magnification that exaggerates the slopes of the faces 001 and 110. It seems apparent that the 111 plane of the far part of Fig. 4.39 are differently oriented from the near part to the extent of about 25 seconds of arc. The sharp demarkation was probably caused by relative slippage of the two parts caused by temperature gradients, caused in turn by thermal shock in reestablishing contact with the melt after the interruption. It is known that at these temperatures dislocations can move in germanium (i.e., germanium is "plastic" to a certain extent). Figure 4.39*a* also shows some lineage—slightly misoriented material near the surface.

Figure 4.40 supplies the details of a double crystal goniometer made for the crystal perfection studies just described. X-rays enter (large arrow at upper left), strike the first crystal at A, reflect onto crystal B, and thence are reflected into the counting tube. Above A, we have the housing that retains the air scattering of the direct beam. Above this is a rotatable cap that stops air scatter from the once-reflected beam. Crystal A is held by a frame that can be tipped by a small micrometer screw (not shown). The post that carries A can be rotated about a vertical axis and can be set by an angle scale seen at its lower end. This assembly fits over a post that carries a fiduciary mark for the angle scale already mentioned. Crystal B

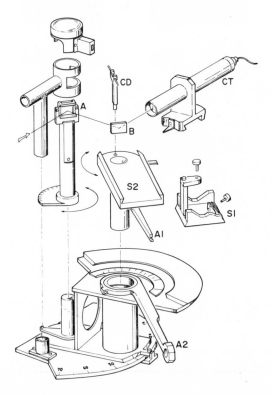

Figure 4.40 This double crystal goniometer, built for crystal perfection studies, can also achieve highest precision orientation.

is carried by a barrel holder such as mentioned in a previous section. The barrel holder goes into the vees of the sled S1 and this is slid along the slide S2 while the crystal pushes the free sliding pin in the centering device CD. When the ring groove in the free pin of CD matches lines on its frame, the crystal is centered.

The counting tube CT slides around on the curved dovetail to be set at angles 2θ as indicated by the angle scale just inside the curved dovetail. The slide S2 is carried on a shaft that goes into ball bearings in the lower assembly. An arm A1 extends from the slide assembly and goes into a position near arm A2. Arm A2 carries a large drum micrometer that bears against arm A1 and gives a means of turning the slide assembly through small determinable angles. With a micrometer screw axis 9.696 in. from the slide rotation axis and with one division of the micrometer representing 0.0002 in. movement, one division gives 4 seconds rotation to crystal B. Since tenths of a division are readily estimated, the angular control is seen to be adequate. Arm A2 can be set for any range of θ desired. A film holder (not shown) clips onto the slide between crystal B and the counting tube.

The Method of A. R. Lang

Figure 4.41 presents the method A. R. Lang has used to photograph dislocations and other defects in crystals. Here X-rays from a target T pass through a fine defining slit S, thence to a thin crystal X. Internal reflection sends part of the beam onto a photographic film F, while the

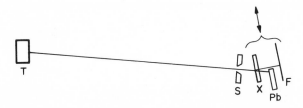

Figure 4.41   The system of A. R. Lang for photographing dislocations in thin crystal plates.

transmitted beam is stopped by a lead shield Pb. The crystal X and the film F move together in the direction of the arrow. The track on which they move must be very straight. An adjustment is provided to set the Bragg angle to achieve maximum reflection. The sliding motion of X and F can be either a very slow translation or an oscillating motion.

A GEIGER TUBE COUNTING RATE CIRCUIT

For the absolute determination of X-ray intensities necessary to identify crystal structures, proportional counters or scintillation counters are highly desirable. However crystal orientation merely requires finding reflection maxima, and Geiger tubes do this adequately with simpler circuitry than that required for the other two. It is not always convenient to buy a counting rate circuit to go with the Geiger tube, hence we reproduce schematically an integrated circuit devised by Leslie Yingst of Stanford University Microwave Laboratory (Fig. 4.42).

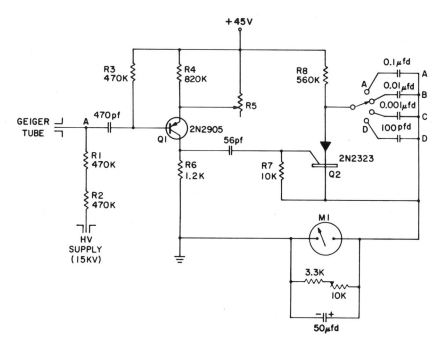

Figure 4.42 Integrated circuit devised by Leslie Yingst, Stanford University Microwave Laboratory.

A word about Geiger tubes. The tube is filled with a gas that is proper for the radiation being used, although it may not be very sensitive for other wavelengths. A very thin window allows the radiation to enter but keeps the gas from escaping. There is also a "quencher" in the tube. The quencher ends the electric breakdown caused by a photon and makes the tube ready for another photon. Otherwise the first photon would start an avalanche breakdown and the tube would not recover. Hence the tube could not be used to measures rates of photon arrival. The quencher is slowly used up. Allowing the direct X-ray to enter the tube quickly depletes it, however. A copper foil 0.10 mm thick should be placed over the tube if it is necessary to locate the tube in the direct beam to check the zero error of the graduated circle used for setting the tube at proper Bragg angles.

5

Optical Methods of Crystal Orientation

OPTICAL ORIENTATION

Although X-ray orientation is the most accurate method for establishing orientation, in some cases optical methods are quicker and involve cheaper equipment that is safer to use. Optical methods also can be used to simplify subsequent X-ray orientation as mentioned in Chapter 4.

There are two approaches to optical orientation: those using reflection and those using double refraction; we include under reflection methods some cases in which light is transmitted through the crystal, although the notion of double refraction is not involved. The reflection methods are simpler in concept and more general in application when natural faces, cleavage faces, or etch pits can be employed. Double refraction methods give no help on cubic crystals, partial orientation only for uniaxial crystals, and complete orientation for biaxial crystals—orthorhombic, monoclinic, and triclinic crystals. Reflection methods can be applied to all classes of crystal.

Reflection Methods

A good illustration of a reflection method is that used on ammonium dihydrogen phosphate (ADP) crystals for making sonar systems during World War II. Grown from a saturated water solution, most of these crystals were about 2 in. square by 10 in. long. Pyramidal faces were well developed and highly reflecting, but the four prism faces were tapered, hence inaccurate. ADP is tetragonal, space group $I\bar{4}2d$, with $a_0 = 7.499$ Å, $c_0 = 7.548$ Å. Since the pyramidal faces are (101) and (011), their normals make an angle $\tan^{-1} 7.548/7.499 = 45°11'$ with the z axis.

The orienting equipment contained three lenses (see Fig. 5.1). The first lens sent a parallel beam of light from a small filament lamp along what

94

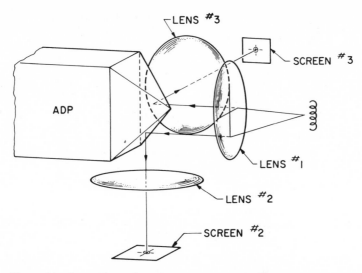

Figure 5.1   A complete orientation by light reflection from natural faces.

was to be the c axis of the crystal. The four pyramidal faces were thus covered with this light, which meant that they reflected four parallel beams. The lower pyramidal face sent a parallel beam nearly straight down. This was caught by lens number 2 and focused on screen number 2. The back pyramidal face (Fig. 5.1) sent a beam through lens number 3 to be focused on screen number 3. A piece of wood resting on a platform supported the crystal. A guide strip on the platform kept the edge of the wood parallel with the eventual c axis direction. The wood carried three lumps of plasticine on which the crystal was to rest. The plasticine was then surrounded by a freshly mixed quick-setting plastic (Norace Cement, made by the Norton Company). The crystal was placed on the plasticine and pressed down into the unset plastic while an investigator watched the two screens. When both images of the lamp filament were centered on the cross lines within the limiting circles, the board was removed from the platform and laid aside. In 20 minutes the plastic was hard enough that the crystal could be sawed into slices perpendicular to the c axis. The plasticine held the crystal in place while the plastic was hardening.

A second illustration of reflection methods is that of reflection from the walls of etch pits. The system illustrated in Fig. 5.2 gives rough orientation for quartz slabs sawed perpendicular to the c axis. Etch pits on (00·1) planes of quartz are seen under the microscope as triangular pits. Reflection from the walls of these pits show a triangular figure and three

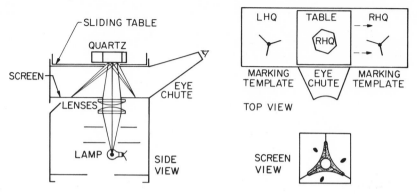

Figure 5.2   Approximate orientation by means of transmitted light scattered from etch pits.

spots. From this one can pencil onto the slab, through a template, the direction of the $+X$ axes for either right or left quartz.

By transmitting the light through the slab, even better figures can be made to appear on properly etched surfaces, as indicated in Fig. 5.3 for etched sections of cross-cut quartz.

PASSAGE OF LIGHT THROUGH CRYSTALS

In well-annealed glass, light has the same velocity for all directions of propagation. The same is true for cubic crystals but not for crystals of any other class. This complicates the problem of light transmissions through such crystals, because for crystals in which light velocity depends on the direction of propagation there are in general two velocities for every direction of propagation; that is, these crystals are "birefringent." Unpolarized light entering such a crystal breaks up into two equal components whose electric displacements are mutually perpendicular (i.e., they are polarized at right angles to each other). These components travel through the crystal with different velocities.

For a given crystal at a given temperature and for a given wavelength of light, we can associate an ellipsoid that describes the passage of light through the crystal. This ellipsoid is called the "indicatrix." Its longest axis is the largest refractive index, its shortest axis is the smallest refractive index. The third axis is equal to the intermediate index. We imagine a plane (Fig. 5.4) perpendicular to any arbitrary direction of propagation, the plane passing through the center of the ellipsoid. This plane intersects the surface of the ellipsoid in an ellipse. The light breaks up into two components with electric displacement directions, one along

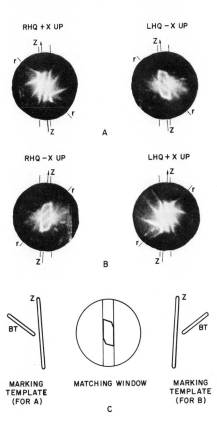

RHQ + X UP   LHQ − X UP

A

RHQ − X UP   LHQ + X UP

B

MARKING TEMPLATE (FOR A)   MATCHING WINDOW   MARKING TEMPLATE (FOR B)

C

Figure 5.3 Some of the patterns revealed by the equipment of Fig. 5.2.

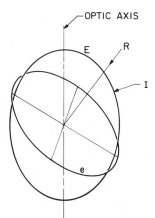

OPTIC AXIS

Figure 5.4 The optical indicatrix for a noncubic crystal.

the major axis of the ellipse (the slower component), the other along the minor axis (the faster component). The indices of the two components are proportional to the lengths of the ellipse axes.

The ellipsoid must satisfy the symmetry of the crystal. Hence for cubic crystals the ellipsoid is a sphere, and a plane perpendicular to any direction of propagation cuts the sphere in a circle. A circle has no definite major or minor axes: the beam does not break into components.

Uniaxial Crystals

A uniaxial crystal has an indicatrix that is an ellipsoid of revolution, that is, one made by rotating a generating ellipse about its major or minor axis. The rotation produces a prolate or an oblate ellipsoid or spheroid, which divides uniaxial crystals into two types, uniaxial positive (prolate ellipsoid) and uniaxial negative. The axis of rotation of the generating ellipse is called the optic axis. Light traveling along the optic axis does not break into components because a section of the ellipsoid is a circle if the section is perpendicular to the optic axis.

Perpendicular to the rotation axis is the "rotating axis." This axial length is common to all ellipses cut from the spheroid by planes that pass through the center of the spheroid. Hence for all directions of propagation one velocity is the same for all; the other velocity varies with direction. The component not changing with direction is called the ordinary component, the other is termed the extraordinary component. The refractive indices are written N_ω for the ordinary component and N_ε for the extraordinary component. If N_ε is greater than N_ω, the crystal is optically positive; otherwise it is optically negative.

Wave Normals, Rays, and Polarizers

Consider a plane parallel light wave. At any given instant and at any arbitrary point z in the wave, all points with phase equal to that at z lie in a plane called a wave front. The normal to this plane is called the wave normal. If the wave velocity changes appreciably for a small change in wave normal direction, the energy flow is not along the wave normal; instead the energy "walks off" a bit to one side along another direction, the ray direction. This allows us to separate the two components into which the crystal resolved the original unpolarized light. Polarizing prisms (Nicol, Ahrens, Glan, Thompson, etc.) use this effect to deliver plane polarized light. Also some crystals absorb much more of one component than the other. Polarizing filters use this effect to give almost pure polarized light.

Snell's Law

Consider a light wave normal that strikes a boundary between two media. If the incoming wave normal is perpendicular to the boundary, the departing wave normal is also perpendicular to the boundary. This is not true of "ray directions" unless the ray direction lies along the wave normal. If the incoming wave normal strikes the boundary with an angle of incidence i, the departing wave normal has an angle of refraction r, where

$$N_i \sin i = N_r \sin r \qquad (5.1)$$

where N_i is the refractive index for the incoming wave and N_r is the index for the refracted wave.

For light passing between two isotropic media, Snell's law (Eq. 5.1) permits the easy calculation of the refraction angle r if we know N_i, N_r, and i, the angle of incidence. This is possible because N_i and N_r do not change with direction. In addition, Snell's law can be used for the ordinary wave of a uniaxial crystal but not for the extraordinary wave, since N_ε varies with direction. In biaxial crystals there is no "ordinary" wave, and the refraction of both components is complicated by variation of index with direction.

Biaxial Crystals

For biaxial crystals (orthorhombic, monoclinic, and triclinic classes) the indicatrix is a triaxial ellipsoid, not one of revolution. Its longest axis is N_γ, where N_γ is the greatest refractive index and its shortest axis is N_α, the smallest index. The intermediate index N_β is independent of N_γ and N_α.

Given the direction of the electric displacement of a light component in a given crystal, the velocity, hence the index, are fixed: all directions of wave propagation that have the same displacement direction have the same velocity. The direction of the electric displacement for maximum velocity is called X, that for the minimum is called Z, and Y is perpendicular to both Z and X, whereas Z and X are mutually perpendicular. The directions X, Y, and Z have no necessary relation to the x, y, and z axes we defined in terms of the axes a, b, and c, except that these directions must satisfy the symmetry conditions for their crystal class.

A triaxial ellipsoid has two circular central sections. The normals to the two circles are called the optic axes. For wave normals in these two directions, beams are not broken into components. Since there are two optic axes, the crystals are called biaxial. For these two directions the refractive index is N_β. The angle between one of these and Z is called V.

Here

$$\tan^2 V = \frac{1/N_\alpha^2 - 1/N_\beta^2}{1/N_\beta^2 - 1/N_\gamma^2} \qquad (5.2)$$

The Y direction is always perpendicular to the plane of the optic axes.

Biaxial crystals are also classified as optically positive ($V < 45°$) and optically negative ($V > 45°$). For optically positive crystals, Z is the acute bisectrix. Under the polarizing microscope birefringent crystals can be identified as uniaxial or biaxial, optically (+) or optically (−). These are diagnostic features useful for crystal identification. Since the indicatrix must have at least the symmetry of the crystal system, we see that in orthorhombic crystals X, Y, and Z must fall along a, b, and c, but not necessarily in that order. For example, diaspore is orthorhombic, the optic plane is 010, and $Z = a$; the mineral is optically (+) with $2V = 84°20'$ for yellow light. In diaspore the optic angle is smaller for red light than for blue light. All these details are helpful in identifying diaspore under the microscope. Often we are given $2E$ but not $2V$. The angle observed outside the crystal (i.e., $2E$) can be corrected for refraction by Snell's law. Here the effective refractive index is N_β. Immersion in a liquid of index N_β allows us to measure $2V$ directly.

With monoclinic crystals the b axis must be one of X, Y, or Z. The remaining two can fall anywhere in the 010 plane. For example, colemanite is optically (+) and the acute bisectrix Z lies in the (010) plane between c and a, making an angle of 83° with the c axis. Also $X = b$ and Y is 7° back of the c axis in the 010 plane. Both values depend on the color of light and the temperature of the crystal.

For triclinic crystals X, Y, and Z may fall anywhere (but must be mutually perpendicular). The directions of X, Y, and Z may change wildly with the color of the light and the temperature of the crystal.

From the data just presented, we realize that biaxial crystals can be completely oriented optically. However there is one difficulty: the directions of X, Y, and Z, as well as the value of V, are sometimes altered by impurities.

PASSAGE OF LIGHT THROUGH A UNIAXIAL BASAL PLATE

A "basal" plate is one whose two major faces are perpendicular to the c axis. Given the previous description of how light breaks up into components while traveling through a crystal, we see that light with its wave normal directed along k (Fig. 5.5) breaks into two components. One, the ordinary component, has its electric displacement direction along D_ω, where D_ω is perpendicular to k and c. The other component has its

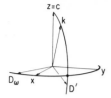

Figure 5.5 Light with wave normal directed along k breaking into two components.

electric displacement along D'' where D'' is perpendicular to k and D_ω. That is, as seen from above, one electric displacement direction is radial, the other tangential, as illustrated for many directions k in Fig. 5.6. For uniaxial *positive* crystals the slow component has its electric displacement in the radial direction; the fast component is the ordinary component and is tangential.

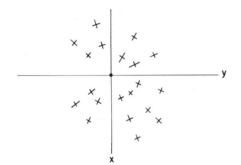

Figure 5.6 The extinction directions for several directions of k.

We consider now polarized light entering and passing normally through any crystal plate, an optic axis *not* being perpendicular to the plate faces. In Fig. 5.7 we assume that the incident light is polarized along y (i.e., its electric displacement direction is along y). On breaking into components

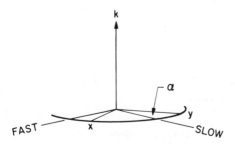

Figure 5.7 Light polarized along y breaks into a slow and a fast component.

here, the slow component will have an electric displacement of amplitude

$$D_s = D_0 \cos \alpha \qquad (5.3)$$

and the fast component will have amplitude

$$D_f = D_0 \sin \alpha \qquad (5.4)$$

where D_0 is the electric displacement amplitude of the incident light. In Fig. 5.8 light from an extended source passes through a polarizer, through a crystal, through another polarizer, and finally to an eye. The upper polarizer (generally called the analyzer) is set so that in the absence of a crystal, no light reaches the eye from the source: the polarizers are "crossed." In the presence of the crystal, light beams from different parts

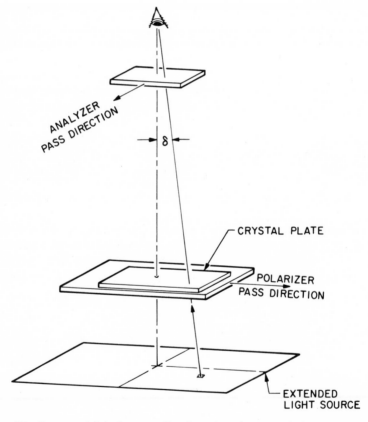

Figure 5.8 Passage of light in many directions through a crystal plate placed between crossed polarizers.

of the field are differently decomposed and are of different amplitudes. Just as the light enters the crystal the two components are of equal phase, although usually of unequal amplitudes. As the components progress through the crystal with unequal velocities, the phases get more and more out of step. At a path length l in the crystal there are lN_ω/λ_0 ordinary waves and lN''/λ_0 extraordinary waves. Here λ_0 is the wavelength of the light in free space. Hence the phase difference is

$$(N'' - N_\omega)\frac{l}{\lambda_0} \quad \text{cycles} \tag{5.5}$$

Figure 5.9 shows the result of adding two light components that are polarized at 90° to each other and are of the same phase and amplitude. The result is plane polarized light, and it can be stopped by a polarizer set to extinguish light polarized in this direction. In Fig. 5.10 there is a phase difference of 30°, and the result clearly is not plane polarized light but elliptically polarized light; moreover, there is no direction in which the extinction direction of the polarizer could be set to stop the light. If the phase difference is 90° and the amplitudes are equal, the resultant electric displacement sweeps out a circle: it is "circularly polarized" light. Only part of this light will be passed by an analyzer set to extinguish light with zero phase difference between its components. If the phase difference is 180°, as in Fig. 5.11, we again have plane polarized light, but the light is polarized at right angles to that for a phase difference zero. Hence if the analyzer is set to stop light of zero phase difference between the two

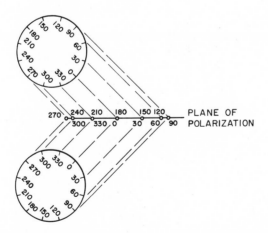

Figure 5.9 How phase differences are caused by the different velocities. Here plane polarized light is produced and can be extinguished by a properly set analyzer.

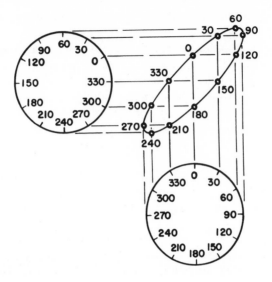

Figure 5.10 The phase difference causes elliptically polarized light; no orientation of an analyzer can extinguish the light.

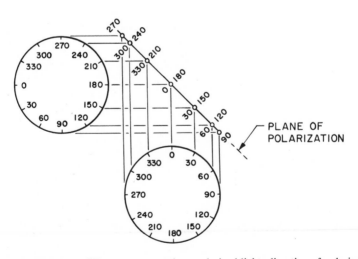

Figure 5.11 The phase difference causes plane polarized light; direction of polarization is 90° different from that of Fig. 5.9.

components, it passes completely light whose phase difference is 180°.

For a given thickness t of crystal, as the wave normal departs further from the c axis direction, two effects combine to make the phase difference between the two components increasingly great: the difference $(N'' - N_\omega)$ increases with angle, and the path length l inside the crystal increases with angle δ (Fig. 5.8). The value of $(N'' - N_\omega)$ increases roughly as the square of the sine of δ.

The net result of the circumstances just outlined is that the eye of Fig. 5.8 sees a bull's-eye ring pattern crossed with a dark cross, as in Fig. 5.12.

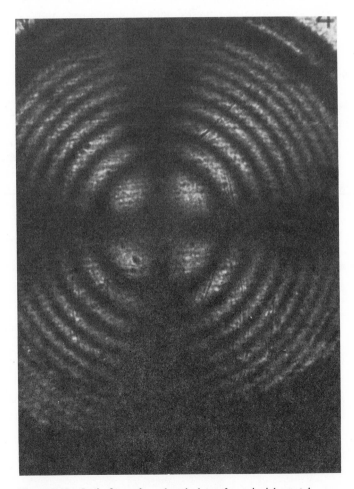

Figure 5.12 Optic figure for a basal plate of a uniaxial crystal.

The center of the bull's-eye is the optic axis. The two arms of the dark cross are the directions of the polarizer and the analyzer extinction directions. If the crystal has a screw axis n_m, where $n \neq 2m$, it is "optically active" and can rotate the plane of polarization of plane polarized light. This effect destroys the cross near the optic axis. A basal plate of quartz rotates yellow light $21.7°$ per millimeter of thickness. This rotation falls off rapidly as the angle between light path and the optic axis increases.

If the major faces of the crystal plate are not accurately perpendicular to the optic axis, the bull's-eye will be off center; it can be recentered, however, by turning the plate about a transverse axis. If the plate is immersed in an immersion fluid having a refractive index that matches N_ω of the crystal, the amount the crystal must be turned to center the pattern is a true measure of the departure of the plate normal from the c axis. If the immersion fluid does not match the ordinary index of the uniaxial plate, a refraction correction may be made.

It should be emphasized that this bull's-eye pattern (called an "optic figure") tells us only the direction of the optic axes, and in the case of optically active crystals, the "handedness" may be observed: if the analyzer is rotated clockwise, the rings expand for right-hand quartz. Both N_ω and N_ε change with the wavelength of the light; thus if white light is used, only a few rings may be seen because of color overlapping. A great many rings may become visible if monochromatic light is employed. Also a pattern can appear transverse to the optic axis, as in Fig. 5.13. For a nonoptically active crystal it is of some importance to know the angular radius of the inner ring. If we cannot see the inner ring, it is difficult to center the pattern. Light with its wave normal at an angle θ from the optic axis has refractive indices N_ω for the ordinary beam and N'' for the extraordinary. Here

$$\left(\frac{1}{N_\omega''}\right)^2 = \frac{\cos^2 \theta}{N_\omega^2} + \frac{\sin^2 \theta}{N_\varepsilon^2}$$

From this we deduce that the birefringence for this direction is approximately

$$B = \frac{N_\omega}{N_\varepsilon}\left(1 + \frac{N_\omega}{N_\varepsilon}\right)\frac{B_0}{2} \sin^2 \theta$$

where $B_0 = N_\varepsilon - N_\omega$. If N_ε and N_ω are not too different, this expression reduces to $B = B_0 \sin^2 \theta$. The path length at angle θ in a plate of thickness t is $l = t \sec \theta$. Hence by Eq. 5.5 the phase difference for the Nth ring is $N = Bl/\lambda_0$. Hence

$$\sin \theta \sim \sqrt{\frac{N\lambda_0}{tB_0}}$$

and the thinner the crystal, the larger the inner ring.

Figure 5.13   Uniaxial plate viewed perpendicularly to the optic axis with monochromatic light.

PASSAGE OF POLARIZED LIGHT THROUGH BIAXIAL PLATES

An obvious extension of the preceding analysis explains the appearance of the "optic figure" if a biaxial plate is viewed as in Fig. 5.8. Such an optic figure is presented in Fig. 5.14. To see both axes simultaneously may require a large angular field if the angle $2V$ between the two axes is large. Without the use of an immersion fluid, the external angle $2E$ is even larger—for gypsum, $+ V = 58°$ but $2E = 95°$. However in a fluid of index 1.530, the observed angle is 58°. The observation and measurement of optic figures can be facilitated by the use of a conoscope (i.e., by "conical viewing").

The Conoscope

In the conoscope we take advantage of the principle of Fig. 5.8 but use lenses to give a large angular field from a small light source. The conoscope of Fig. 5.15b has several desirable features: (1) a small monochromatic light source is used to give a large angle of field; (2) an immersion fluid is employed; (3) the focus is unaltered by the immersion fluid, which can have any index including 1.000 (air); (4) the angle between plate normal and optic axis can be measured directly if the immersion

Figure 5.14   Biaxial plate, showing the two optic axes.

fluid index matches the proper crystal index; (5) the analyzer is rotatable, permitting us to establish the handedness of optically active crystals.

A more compact instrument is illustrated in Fig. 5.15a. Instead of a tank full of immersion fluid, this instrument uses only a few drops. The crystal plate is pressed between two glass hemispheres. The upper hemisphere is set into a disk that can be rotated in a ring, and the position angle can be read from a divided circle on the disk. The ring can be tipped as shown and the angle of tip can be read on another scale. For very thin plates observation can be made only through the crystal thickness; indeed, if an optic axis is too steeply inclined (i.e., beyond the critical angle), it cannot be observed without an instrument such as this or the conoscope (Fig. 5.15b). The critical angle in air is $\sin^{-1} 1/N$, where N is the ordinary index for uniaxial crystals, but it is N_β for biaxial crystals. In a quartz plate in which the optic axis is more than $\sin^{-1} 1/1.544 = 40.4°$ from the thickness direction, the optic axis cannot be observed. With glass hemispheres of index N_g, if immersion fluid drops of index $N_f > N$, the critical angle becomes $\sin^{-1} N_g/N$. If $N_g > N$, there is no critical angle. However even hemispheres with $N_g = 1.6$ will enlarge the critical angle of sapphire ($N_\omega = 1.767$) to 65.9° from its value of 34.5° in air.

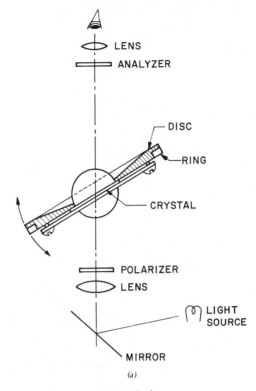

Figure 5.15a   Conoscope. A more compact device.

The Stauroscope

The simple crossed polarizers of Fig. 4.2 can be rendered more precise by a split field device. Unfortunately the extinction range of simple crossed polarizers is rather broad because of the dependence of a minimum of transmitted energy near its turning point, where the energy is approximately proportional to the square of the angle of error. By cutting a wedge from the polarizer and closing the gap, an improved performance is obtained. The eye (Fig. 5.16) now matches the intensities of the two parts by rotating the crystal and reference bar. The departure of the crystal length direction from a slow direction or a fast direction is readily read from the angle scale. To check the stauroscope, we first turn the analyzer, to equalize the light intensity on the two halves of the field with no crystal in place. Next we place a parallel-edged crystal against the reference bar and equalize the light intensity in the two halves of the crystal by turning the reference bar. We then read the angle scale and turn the crystal over,

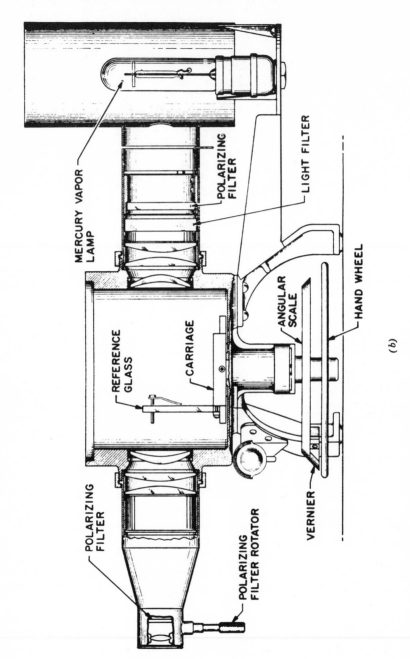

Figure 5.15b Conoscope. Instrument featuring tank of immersion fluid.

(b)

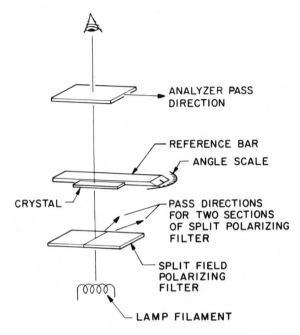

Figure 5.16   A stauroscope.

180° about its length, and rebalance. If the two angle scale readings are not equal but of opposite sign, there is a zero error, which can be eliminated by moving the index mark or, alternatively, rotating the angle scale slightly.

FINDING THE OPTIC AXIS IN A RECTANGULAR BLOCK OF UNIAXIAL CRYSTAL FROM EXTINCTION OBSERVATIONS ALONE

We suppose the optic axis in a rectangular block of uniaxial crystal lies along $[XYZ]$, the coordinate axis being chosen along block edges, as in Fig. 5.17

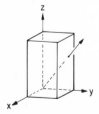

Figure 5.17   Optic axis along $[X, Y, Z]$ but what cone values of x, y and z?

Looking through the crystal between crossed polarizers, we find that for light traveling parallel to the x axis, the direction $[0yz]$ is an extinction direction, but so is the direction of $[0z\bar{y}]$. One of these two directions is perpendicular to the optic axis (the scalar product of $[xyz]$ and $[0z\bar{y}]$ is zero). It is not obvious which of the two directions is perpendicular to the optic axis from extinction observations alone. We draw these two extinction directions in Fig. 5.18, where the measured angle is denoted u; however we allow for two possibilities, u' and u'', one of which is measured u, the other being the negative of its complement. We see that $\tan u' = z/y$, $\tan u'' = -y/z$, u'' being taken as negative. Hence $u' - u'' = 90°$.

Similarly, looking along the y direction we observe two extinction angles v' and v'' with $\tan v' = x/z$ and $\tan v'' = -z/x$, as in Fig. 5.19. Again one of the two extinction directions is perpendicular to the optic axis, but we don't know which one.

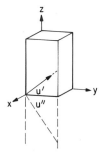

Figure 5.18   Extinction directions as viewed along X.

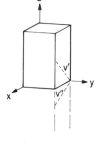

Figure 5.19   Extinction directions as viewed along Y.

Looking along the z axis (Fig. 5.20), the directions w' and w'' are extinction directions, and one is perpendicular to the optic axis. Here $\tan w' = y/x$, $\tan w'' = -x/y$. We now see that $\tan u' \tan v' \tan w' = 1$, and

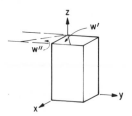

Figure 5.20   Extinction directions as viewed along Z.

$\tan u'' \tan v'' \tan w'' = -1$; other (i.e., mixed) combinations lack this simplicity. These facts allow us to identify the three directions perpendicular to the optic axis and to determine which three are not.

We can now state that

$$\begin{pmatrix} x \\ y \\ z \end{pmatrix} = \begin{pmatrix} \tan v' \\ \cot u' \\ 1 \end{pmatrix} \times \text{a constant} = \begin{pmatrix} \cot v'' \\ \tan u'' \\ -1 \end{pmatrix} \times \text{a constant}$$

As an example, we take the set $u = 25°$, $v = 52°$, $w = 70°$, which gives

$$\tan u = 0.466 \quad \text{or} \quad -2.145$$
$$\tan v = 1.28 \quad \text{or} \quad -0.781$$
$$\tan w = 2.747 \quad \text{or} \quad -0.364$$

To cover half the permissible combinations, we write the multiplication net:

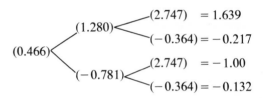

$$
\begin{aligned}
&(0.466) \begin{cases}
(1.280) \begin{cases}
(2.747) &= 1.639 \\
(-0.364) &= -0.217
\end{cases} \\
(-0.781) \begin{cases}
(2.747) &= -1.00 \\
(-0.364) &= -0.132
\end{cases}
\end{cases}
\end{aligned}
$$

This indicates that $\tan u'' = 0.466$, $\tan v'' = -0.781$, $\tan w'' = 2.747$, whence $[xyz]$ lies along the direction $[-1/0.781, 0.466, -1]$ or more simply, along $[1.28, \overline{0.466}, 1]$.

We do not need to form the other half of the permissible combinations because we now have the answer. If we had started with -2.145, the calculation net would have been

$$
\begin{aligned}
&(-2.145) \begin{cases}
(1.280) \begin{cases}
(2.747) &= -2.051 \\
(-0.364) &= 0.999
\end{cases} \\
(-0.781) \begin{cases}
(2.747) &= 4.602 \\
(-0.364) &= -0.610
\end{cases}
\end{cases}
\end{aligned}
$$

Hence $\tan u' = -2.145$, $\tan v' = 1.280$, $\tan w' = -0.364$, and

$$[x, y, z] = \left[1.280, \frac{-1}{2.145}, 1 \right] \quad \text{or} \quad [1.280, \overline{0.466}, 1]$$

If two faces each have zero and 90° extinction, the optic axis is parallel to the face having inclined extinctions. The optic axis is parallel to *one* of the two inclined extinction directions—but which one? Our analysis breaks down because the multiplication nets involve multiplication of zero and infinity; the result is indeterminate. A quick X-ray check will tell which of the two inclined extinction directions is the optic axis.

We have assumed up to now that the specimen is a rectangular parallelopiped. However extinction directions can be observed with cylinders. Although it is natural to pick the cylinder axis as the z axis and two directions of noninclined extinction as the x and y axes, this approach leads us into the indeterminacy trap just mentioned. We can pick z along the cylinder axis but choose x and y 30°, say, from the directions of noninclined extinction, thus avoiding the trap.

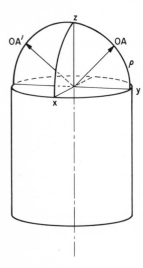

Figure 5.21 Extinction directions in a cylinder.

In Fig. 5.21 we illustrate the situation. Viewing parallel to the cylinder axis between crossed polarizers, we get parallel extinction relative to the indicated x and y axes. Viewing parallel to the y axis we get parallel extinction, but viewing along the x axis we find inclined extinction with an angle ρ. We do not know that the optic axis lies along OA, however; it might just as well be along OA', 90° away.

If the optic axis makes an angle ρ with the y axis as in Fig. 5.21, we describe the optic axis as along the vector $\begin{pmatrix} 0 \\ \cos \rho \\ \sin \rho \end{pmatrix}$. We now rotate the cylinder 30° clockwise about the cylinder axis, and the optic axis is now along

$$
\begin{pmatrix}
\dfrac{\sqrt{3}}{2} & \dfrac{1}{2} & 0 \\[2mm]
-\dfrac{1}{2} & \dfrac{\sqrt{3}}{2} & 0 \\[2mm]
0 & 0 & 1
\end{pmatrix}
\begin{pmatrix}
0 \\[2mm]
\cos\rho \\[2mm]
\sin\rho
\end{pmatrix}
=
\begin{pmatrix}
\dfrac{1}{2}\cos\rho \\[2mm]
\dfrac{\sqrt{3}}{2}\cos\rho \\[2mm]
\sin\rho
\end{pmatrix}
$$

We should now read $\tan u = z/y = 2/\sqrt{3}\tan\rho$, $-\sqrt{3}/2\cot\rho$ $\tan v = x/z = \frac{1}{2}\cot\rho$, $-2\tan\rho$, $\tan w = y/x = \sqrt{3}$, or $-1/\sqrt{3}$; a multiplication net now reads

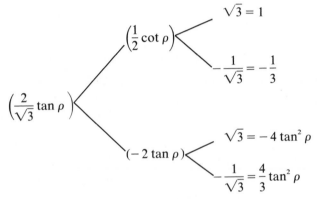

and shows that we may take $\tan u' = 2/\sqrt{3}\tan\rho$, $\tan v' = \frac{1}{2}\cot\rho$. Hence

$$
\begin{pmatrix}
x \\[2mm]
y \\[2mm]
z
\end{pmatrix}
=
\begin{pmatrix}
\tan u' \\[2mm]
\cot u' \\[2mm]
1
\end{pmatrix}
=
\begin{pmatrix}
\dfrac{1}{2}\cot\rho \\[2mm]
\dfrac{\sqrt{3}}{2}\cot\rho \\[2mm]
1
\end{pmatrix}
$$

Multiplying this by $\sin\rho$ to normalize it, we have

$$
\begin{pmatrix}
x \\[2mm]
y \\[2mm]
z
\end{pmatrix}
=
\begin{pmatrix}
\dfrac{1}{2}\cos\rho \\[2mm]
\dfrac{\sqrt{3}}{2}\cos\rho \\[2mm]
\sin\rho
\end{pmatrix}
$$

Rotating this 30° counterclockwise yields

$$
\begin{pmatrix} \dfrac{\sqrt{3}}{2} & -\dfrac{1}{2} & 0 \\[2ex] \dfrac{1}{2} & \dfrac{\sqrt{3}}{2} & 0 \\[2ex] 0 & 0 & 1 \end{pmatrix} \begin{pmatrix} \dfrac{1}{2}\cos\rho \\[2ex] \dfrac{\sqrt{3}}{2}\cos\rho \\[2ex] \sin\rho \end{pmatrix} = \begin{pmatrix} 0 \\[2ex] \cos\rho \\[2ex] \sin\rho \end{pmatrix}
$$

which is the direction of the optic axis on the original coordinate axes.

If the edges of the rectangular parallelopiped are slightly rounded and smooth, extinction can be obtained through corners, to resolve the indeterminacy.

"COMPLETE" OPTICAL ORIENTATION OF BIAXIAL CRYSTALS

The complete orientation of biaxial crystals (orthorhombic, monoclinic, triclinic) is possible but not practiced in industry to the author's knowledge. In the case of monoclinic and orthorhombic crystals, the "complete" has a certain ambiguity that is readily removed.

Any device such as the conoscope that can observe the optic figures can be used to confirm adjustment of the crystal relative to a base plate. In the simplest such scheme, we make both optic axes parallel to the base plate, with the acute bisectrix along the length of the plate. The ambiguity is that the optic axis configuration has three binary axes and three planes of symmetry, whereas the crystal itself probably does not have these symmetry elements. Of the four positions (Fig. 5.22), only one may be correct. For a triclinic crystal the use of more than one wavelength of light may remove the ambiguity because the optic axes shift with light

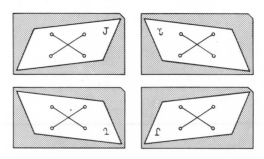

Figure 5.22 The ambiguity of complete optical orientation of a biaxial crystal.

wavelength, and this may indicate the proper choice. However several other simple tests may remove the ambiguity. Observation of natural faces, etch pits, and so on, may resolve the ambiguity. Piezoelectric tests have great value for crystals of low symmetry. A sensitive vacuum tube volt meter (or Heintzen's integrated circuit, Fig. 1.43) can show what directions in the crystal become electrically positive on local pressure.

A similar ambiguity exists for X-ray orientation. We resolved this by noting the angular distance to neighboring atomic planes. The piezo test is also quite valuable here.

A further complication is that of finding the relation between the directions of the optic axes and those of the crystallographic axes a, b, c. There is no problem if this information is contained in the crystal data, but commonly the description is given in terms of the "directions of vibration" for the slow ray and for the fast ray. These vibrations are called Z and X, respectively, but they can have any direction in the crystal, with the limitations that they must be mutually perpendicular and must satisfy the symmetry conditions of the crystal class. Direction of vibration here means the direction of the alternating electric current displacement associated with the electric field of the light wave. The refractive index is uniquely determined by this direction. The direction of propagation may be any direction perpendicular to the vibration direction. The index of refraction for vibration along Z is called n_γ (or sometimes γ) and the index for vibration along X is called n_α (or α). Of course, $n_\gamma > n_\alpha$. A direction Y perpendicular to Z and X is called the intermediate axis, and vibrations along Y have the refractive index n_β (or β). In triclinic crystals α, β, and γ, we define the angles between the crystallographic axes; hence if n_α is called α, n_β, β, and so on, confusion may follow.

Although for some orthorhombic crystals Z lies along c and X lies along a (as in topaz), for others this is not the case. Barite ($BaSO_4$) is orthorhombic with $a_0 = 8.85$ Å, $b_0 = 5.44$ Å, and $c_0 = 7.13$ Å. However a text on mineralogy gives the following information for barite:

$$n_\alpha = 1.636, n_\beta = 1.637, n_\gamma = 1.648$$
$$2V = 36°, \text{opt} (+), a = Z, b = Y, c = X$$

Here $2V$ is the angle between the optic axes but it is the angle "in the crystal." Refraction makes this angle appear different outside unless the crystal is immersed in a fluid of index n_β, the intermediate index of the crystal. If we are given n_α, n_β, and n_γ, we may use the following equation

$$\tan^2 V = \frac{1/n_\alpha^2 - 1/n_\beta^2}{1/n_\beta^2 - 1/n_\gamma^2}$$

to compute, where V is the angle between either optic axis and the direction Z. If V is less than 45° the crystal is optically positive, otherwise it is optically negative.

Consider a crystal plate being viewed with the apparatus of Fig. 5.8. It is assumed that the faces of the plate are polished sufficiently to be transparent and that the plate is untwinned. An index matching fluid and glass cover can substitute for a polish. As the crystal is rotated in its own plane, it may appear brighter, then darker, then brighter, with four dark positions per turn of the crystal. If this is the case, the crystal is birefringent. In the dark positions (i.e., the so-called extinction positions), directions in the crystal parallel to the pass directions of the analyzer and polarizer are called extinction directions of the crystal. The optic axis of a uniaxial crystal is an extinction direction. The X, Y, and Z directions of biaxial crystals are extinction directions. This implies that the a, b, and c

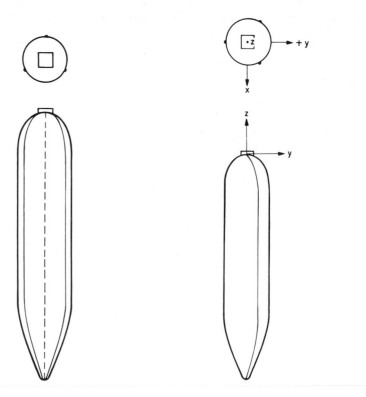

Figure 5.23 Ridges along the boule of a lithium niobate crystal grown with length along $c = z$.

Figure 5.24 Position of axes of crystal in Fig. 5.23.

axes of orthorhombic crystals are extinction directions and that the b axis of a monoclinic crystal is an extinction direction.

If the plate does not show extinction directions, there are two possibilities: the plate is equally dark for all positions or equally bright for all positions. If the plate is dark in all positions, either it is not birefringent or we are viewing in a direction that is not birefringent, an optic axis direction. If the plate is equally bright in all directions it is optically active—quartz, for example, or bismuth germanium oxide (BGO). With an optically active uniaxial crystal such as quartz, slight tipping may be enough to show parts of interference rings: the thinner the plate, the more tipping is required.

The rotation of optically active crystals is dependent on the wavelength; hence with the pass directions of analyzer and polarizer crossed, the crystal appears to be colored. If the analyzer is rotated, the color changes through the spectrum. Extinction can still occur for monochromatic light, but the analyzer and polarizer are seldom at right angles at extinction. Optically active cubic crystals such as BGO, sodium chlorate, and sodium bromate, do not reveal interference rings on tipping. Biaxial crystals can show optical activity also, but it is difficult to observe.

We should not leave the subject of optical orientation without a comment on rough orientation of boules by inspection. Natural crystals generally have natural faces indicating orientation, and boules pulled from a melt often have rudimentary faces and sometimes ridges that are evidence of the crystal symmetry. For example, lithium niobate grown with the length along $c = z$ often has ridges along the length of the boule (see Fig. 5.23). The yz plane is a plane of reflection symmetry, and the direction between boule center and ridge is positive on compression as shown by the polarity tester of Fig. 1.43. Hence since d_{22} is said to be positive (Chapter 13) and expansion, not compression, is considered to be positive, the $+y$ axis must lie midway between two ridges. Now d_{33} is also observed to be positive. Hence the axes are as shown in Fig. 5.24, and this agrees with the rule $-h + k + l = 3n$.

6

Materials

Hard materials may be ground on hard laps; a very hard cast iron can put a polish on diamond. Cast iron, which is hard but still machinable, is ideal for general roughing. Hardened tool steel would be even better for many jobs, but hardening generally warps the surface. For odd jobs on softer crystals brass works quite well. Copper charged with diamond dust is also very effective on hard crystals. Softer crystals can be polished by means of laps made of tin, solder, or lead. These materials give nice finishes if the laps are charged by grinding and the crystal is worked after the lap has been cleaned and the debris of grinding removed by flooding with water or other coolant. Sometimes the best polish comes with working with the lap almost dry.

COOLANTS

One of the desired properties in a coolant is low viscosity, and water is often used as a coolant. However each combination of lap, crystal, and abrasive seems to do best with a particular coolant that can only be chosen after experimenting. Water evaporates rather quickly, but the addition of glycerine or ethylene glycol helps in this respect. Also propylene glycol has been used, and its vapor is safer to breath. A saw can broadcast quite a mist. Fortunately sodium carbonate or even soap can help to eliminate the rusting promoted by water. Water-soluble oils used in machining metals are frequently selected as coolants. Kerosene is used, as well as alcohols such as propyl alcohol. Another cooling fluid, triethanolanime, evaporates much less quickly than water and is infinitely soluble in water.

ABRASIVES

Since abrasion requires abrasives, we now consider abrasive substances. Most of our work is done with one or more of four abrasive materials: aluminum oxide (Al_2O_3), silicon carbide (SiC), boron carbide (B_4C), and diamond (C). We discuss these in turn, leaving polishing materials until later.

Aluminum Oxide

Emery is an impure, naturally occurring aluminum oxide that has been used as an abrasive for centuries. It is a little softer than pure aluminum oxide. Commercial aluminum oxide abrasive has more uniform quality than emery, and most of it is made by fusing bauxite in an electric furnace. Trade names are alundum and aloxite.

Silicon Carbide

Silicon carbide, another product of the electric furnace crystallizes out when sand and coke are fused. The crystals are harder than aluminum oxide but a little more brittle. Trade names are carborundum and crystolon.

Boron Carbide

Boron carbide is also a product of the electric furnace. It is harder than silicon carbide but more expensive. However when cutting very hard materials, the greater hardness is a distinct advantage.

Diamond

Diamond is much harder than any of the previously mentioned abrasives. It is so expensive that it is used sparingly, but on very hard materials it is almost indispensable. Diamond saws are a practical necessity for sawing even moderately hard crystals. Synthetic diamond abrasives are now made in large quantities.

Other Abrasives

Many other abrasives are used commercially but are unimportant in our work. The list includes crushed glass, crushed steel, sand, garnet, and flint. In polishing we occasionally find use for some abrasives never used for gross roughing. These are cerium oxide and chromium oxide, magnesium oxide and tin oxide. At one time iron oxide (rouge) was used extensively, but cerium oxide does the same class of work more quickly. In polishing some soft materials (e.g., metals) chrome oxide is useful. A very fine aluminum oxide called "Linde" is excellent for polishing.

Preparation of Abrasives

The manufacturer of abrasives crushes the material and separates it into sizes. Historically the sizes were named by the number of threads per inch of silk screens that were used for sifting the powders. A common series of screens is 10 threads per inch, 12, 14, 16, 20, 24, 30, 36, 46, 54, 60, 70, 80, 90, 100, 120, 150, 180, 220. The material that passes through the 100 screen but not through the 120 is called "100 grit." Sifting does not give perfect separation by sizes, however: some long pieces slide through holes endwise, and other long pieces, though smaller, may lie flat on the screen and fail to pass through. A rough rule is that on the average, size N abrasive is $0.58/N$ in. across but includes particles ranging from $0.45/N$ to $0.75/N$ in. across. Some smaller pieces are also present in smaller quantities. Above size 220 the abrasives are separated by levigation, generally in water. Finer grits settle more slowly, hence coarse material remains when the finer material is decanted off with the water. The sizes are named to continue the series of the silk sieve separation and obey the formula just given for size versus number. The numbers are 240, 280, 320, 400, 500, 600, and some suppliers continue the series on to 3200.

It is more convenient to think of the small pieces in terms of their dimensions in (μm). We approximate the sizes as: average $15,000/N\mu$, maximum $= 19,000/N\,\mu$, minimum $= 11,000/N\,\mu$. The American Optical Company, an important supplier of abrasives, labels its sizes by arbitrary catalog numbers: M302, M302$\frac{1}{2}$, M303$\frac{1}{2}$, M304, and M305. These correspond to sizes 400, 600, 800, 1200, 1600, and 3200. Optical workers often designate sizes by settling time. If we apply Stokes' law to emery in a meter depth of water, we find the average size to be about $30/\sqrt{T}\,\mu$, where T is the settling time in minutes.

Quality of Sizing

In *Prism and Lens Making* F. Twyman gives a rule for imperfection of grading: $I = 10(S_{max}/S_{av} - 1)$. He says that if I is 4 or less the grading is good. This implies that the maximum size is not greater than 1.4 times the average size. No screening process used on irregularly shaped particles can produce perfect separation. As we mentioned earlier, an elongated particle will pass through a screen or not, according to how its neighbors cause it to approach the screen. Levigation is also imperfect. Convection currents and Brownian movement influence the falling rate, and shape, also, influences the result. We arrive at the equation size $= 30/\sqrt{T}$ by assuming the particles to be spheres and this is unlikely to be the case—fortunately, since sharp points cut better than rounded ones. However even perfect spheres can cut if they are pressed firmly into the workpiece. If the workpiece is brittle, there will be shattering around the sphere, hence material is removed from the workpiece.

Abrasion as a Process for Removal of Material

A harder material scratches a softer brittle one, thus removing material. If the softer body is not brittle, material may be moved but not detached; hence a scratch is made but no weight loss occurs. In sandpaper many scratching points are cemented to a piece of heavy paper, permitting thousands of scratches to be made simultaneously. Continued stroking removes more and more material. Since the removed material must be carried between the grains, there must be space between the grains. The interstices can become full, impeding further removal of material from the piece being worked. Also some grains break off, and finally cutting proceeds at a very slow rate. The action of emery cloth is the same as that of sandpaper.

Similar action occurs when we sharpen a knife on a whetstone, only the abrasive material is a porous stone; thus as grains break off, underlying grains are exposed and the cutting can continue. Since the stone is porous, new interstices are exposed as the grains break off. Again the interstices may become filled and the cutting is then impeded. If the stone is wet with oil or water, the removed material is less able to stick to the grains and plug the interstices. Without the fluid, the removed material can become very firmly impacted in the spaces between grains. If the grains break off fast enough, however, the impacted material is released. Power-driven grinding stones work like the whetstone, but here water or oil may be essential to prevent the workpiece from becoming overheated.

In all the cases considered thus far the abrasive is secured to a block and wears away material from a surface over which it is sliding. We now take the case of an abrasive that is loose but passes between two surfaces that slide over each other. Here the whole grain is exposed to the action and there are no fixed interstices to become impacted—the removed material joins the free abrasive in passing between the surfaces. Generally one piece is used as a tool to shape the other piece. As one piece slides over the other, the abrasive grains roll and slide, gouging material from tool and work. If the tool is soft some grains become partially buried and then cut, as in the sandpaper action. In fact soft tools (or laps) become "charged" in this way, permitting them to cut when moved over the work even after all loose material has been removed. Hence we see that grinding involves forced penetration and scratching, with either loose grains or bound grains.

The most precise grinding is done with charged tools. The fastest grinding is achieved when loose abrasive is plentifully supplied. However, the grains are larger as they pass in between the two surfaces than they are when they pass out from between the two surfaces. This means that a perfect fit between the two surfaces is not easily obtained because the edges of the pieces are cut more than are the centers, giving rounded edges. This effect can be reduced by cutting a grid of channels into the lap, allowing fresh abrasive to be introduced more centrally and making the lap cut the work more quickly.

Cutting with loose abrasive always requires a fluid. The fluid, through its surface tension, helps hold the abrasive to the surfaces. Without the fluid the abrasive would be easily pushed away and would not pass between work and lap. The fluid also acts as a coolant to reduce local temperatures, which could cause damage. Water is one of the best fluids for cooling, but it has some disadvantages. First, it evaporates rather readily, soon leaving the work dry. This tendency can be alleviated by the addition of a little glycerine. Second, its viscosity is low. A liquid soap increases viscosity and holds the abrasive to its task better. Third, water rusts iron laps. Rust inhibitors can be used to combat this condition. A light oil is sometimes chosen as the fluid.

Speed of Cutting and Abrasion Resistance

When cutting a given substance we need a guide to the best material from which to make laps. The material of a lap should be readily cut to the shape needed—flat for the production of flat surfaces, curved for the production of curved surfaces. The lap should cut into the work as fast as possible, and it should resist change of shape due to abrasion rather well. Another factor to consider is the chargeability of the material. For some purposes we need to charge the lap and then work without loose abrasive. In other cases we wish to avoid charging so that we may go from coarse to finer abrasives by washing off the old and applying the new grit without running the risk that coarse charging will cause scratches in the finer stages.

To aid in selecting lap materials we have tested the speed with which several possible lap materials cut quartz with silicon carbide in water and the resistance of materials to abrasion on cast iron with silicon carbide and water (Tables 6.1 and 6.2).

The desirability of a material as a lap can be measured approximately by the product of these two quantities. Table 6.3 gives this factor.

Table 6.3 does not reflect the qualities of chargeability or ease of preparation. Glass is the least chargeable, tin the most chargeable. Hard tool steel is the most difficult to shape because of warpage during the hardening process. Cast iron is hard yet machinable (diamond is commonly faceted on specially hard cast iron but sawed with copper or bronze saws).

TABLE 6.1 COMPARATIVE CUT-
TING SPEEDS OF LAP MATE-
RIALS

Cast iron	1.0
Cast iron with grids	2.0
Hard tool steel	2.0
Brass	0.6
Glass	1.6
Tin	0.7
Aluminum	0.8

TABLE 6.2 RESISTANCE TO AB-
RASION OF SEVERAL MATE-
RIALS

Cast iron	1.0
Hard tool steel	1.5
Brass	0.34
Aluminum	0.28
Quartz	0.13
Glass	0.066

TABLE 6.3 DESIRABILITY OF
MATERIAL AS A LAP MATERIAL

Cast iron	1.0
Cast iron with grids	3.0
Hard tool steel	4.5
Brass	0.2
Aluminum	0.22
Glass	0.1

A tin lap charged with fine abrasive will put an optical finish on hard crystals and, if used with care, on soft crystals. This is the method by which gem stones are faceted, and it has the following advantage: the edges of the surface are not rounded off. If the facets were made by polishing on pitch or felt, the edges would be slightly rounded. This effect is noticeable on a faceted stone, and it detracts from the appearance of the gem.

There is a popular belief that in abrasive conditions soft materials wear better than hard materials. The data of Tables 6.1 through 6.3 deny this. The truth seems to be that resilient materials wear better than brittle ones. This is substantiated by the data on quartz versus brass. Quartz is much harder than brass but is less resilient, and it abrades between two and three times as readily as does brass.

A lap that is very hard and very smooth might not cut well because the abrasive grains would not find a footing with which to resist the motion of the workpiece. Grids help in this.

CEMENTING

Soft plastic clay is often sufficient to hold a crystal for string sawing. Although a certain amount of sawing and grinding of crystals can be done by merely holding the crystal against the saw or lap, it is safer to cement the crystal to a block that can be grasped or clamped more firmly. This is especially true of small crystals, which would be hard to hold firmly in the fingers. If a piece being sawed is allowed to turn slightly in the fingers, it may be seized by the saw, perhaps resulting in injury to the fingers, breakage of the crystal, and damage to the saw. Hence we assume that cementing precedes sawing as indeed it precedes many grinding processes.

The principal cements are waxes or resins that melt at a reasonable temperature.

Cements

The cement should be quite fluent at a temperature the crystal can withstand. Since cements soften near their melting points, however, low melting temperature cements are weak and must be used with care. High melting temperature cements are strong. For crystals that can withstand the greater heat of high melting temperature cements, much more vigorous sawing or grinding is permissible. Table 6.4 gives a few useful cements with their fluent temperatures and solvents for subsequent clean removal.

It is best not to cement the crystal to be sawed directly to the metal mounting block; rather, a scuff plate—a parallel slab of glass, graphite, wood, or ceramic—should be placed between the crystal and the block. Without the scuff plate it is impossible to saw completely through the crystal without sawing into the metal mounting block. It is undesirable to saw into metal because metal cuts much less readily than a brittle crystal, and a saw is easily ruined by entering the metal. It is often undesirable to leave a crystal cut not quite clearly through because although the crystal

TABLE 6.4

Cement	Fluent temperature (°C)	Soluble in
Mutton tallow	45	Trichlor ethylene
Phenyl salicylate	45	Alcohol
Paraffin	95	Benzene
Beeswax	70	Ether
Phenyl benzoate	70	Alcohol, ether
Stearic acid	70	Ether
Half-and-half beeswax—rosin	120	Ether
Half-and-half beeswax with added plaster to give a plastic range of temperature	120	Ether
Glycol phthalate	140	Acetone
Rosin	150	Alcohol, ether
Bayberry wax	60	Benzene

will break through the rest of the cut, a ragged edge is the result (see Fig. 6.1). If we wish subsequently to cement the slab snugly against a surface, the ragged projections will prevent it.

To reduce the hazard of thermal shock, which might crack the crystal, it is best to place the scuff plate on the mounting block and place the crystal on top, with all pieces at room temperature. If the layer of cement must be of uniform thickness, it is best to clamp the assembly together, because surface tension of the molten cement may "float" the crystal. The orientation may be spoiled if the crystal turns in its own plane relative to the mounting block. Small crystals can be secured by means of a clamping tool (Fig. 6.2) in which a metal plate carries a metal post on which a metal

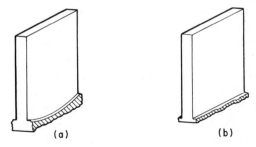

(a)   (b)

Figure 6.1 Result of sawing without a scuff plate.

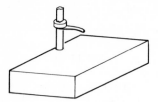

Figure 6.2 Simple clamp to assist in cementing.

collar slides. The collar carries a spring finger. With the mounting plate–scuff plate–crystal assembly on the clamping block, the collar is slid down the post until the spring rests on the crystal. A slight further forced descent of the collar locks the spring tightly against the crystal. The whole assembly is now stable enough to be transferred to the heat source. The danger of thermal shock is small for small crystals, and a hot plate is the most convenient heat source. Thermal shock poses a greater risk for larger crystals, and an oven should be used.

Pieces of cement laid around on the mounting block are watched for signs of melting. When the cement is fluent it is led around the edges of the scuff plate, permitting capillary action to carry it into the space between scuff plate and mounting block and scuff plate and crystal. Grooves in scuff plate and mounting block make for easy entry of the hot cement. Sometimes it is permissible to slide these over each other slightly to aid in the distribution of cement. When the distribution is satisfactory, cooling should be started. For small crystals one can merely transfer the assembly to a large metal plate kept near room temperature. The plate will absorb the heat quickly but gently. For large crystals it is sometimes best to merely turn the oven off and wait for it to cool. If the assembly is removed from the oven and placed on a block of wood, air drafts may cause the crystal to crack. Covering the assembly with a cloth or paper shield after removal may be sufficient to avoid cracking, thus making it possible to keep the oven at constant temperature.

We have used melting cements, not drying cements, because drying cements must lose solvent before they can become hard. Evaporation sometimes occurs only around the edges and it takes an inordinately long time for a drying cement to become firm all the way to the center. Chemicals characterized by a chemical reaction that converts them from liquid to solid do not suffer this drawback, but they cannot be made to let go very readily if the assembly has to be dismantled later. Rochelle salt makes an interesting cement when melted. It can be supercooled to room temperature. Later crystals start to form and the melt solidifies. The process can be expedited by sprinkling a little sodium tartrate or potassium tartrate around the edges to seed the crystal growth. The parts can be separated later by dissolving the cement in water.

Norace Cement, mentioned in connection with Fig. 5.1, was once used for mounting ADP crystals for sawing. This plastic material sets quickly if the accelerator, furfural, is used copiously, it will not melt but separates from crystals rather well, and it can be cut from mounting blocks for reuse of the blocks.

ETCHING

We have mentioned etching as an aid in crystal orientation and in examining for twinning, lineage, and other phenomena. Each crystal is a separate problem and must be approached through chemistry. Also there are two contrasting ends in view: a "smooth" etch or a "crystal" etch. The crystal etch is a slow etch that develops twin boundaries, lineage boundaries, and other characteristics, to reveal crystal imperfections. The smooth etch is a fast etch that leaves a smooth surface and does not reveal twin boundaries and so on.

In etching quartz crystals we use 40% hydrofluoric acid, which cannot be stored in glass bottles. Various solutions of fluorides have also been employed.

A polish etch is given to silicon by a mixture of hydrofluoric acid (3 parts, 48%), nitric acid (5 parts, 70%), with 3 parts glacial acetic acid and 2 parts of aqueous mercuric nitrate (3% solution). Germanium is etched in a mixture of nitric and hydrofluoric acids to which acetic acid and bromine are added. These formulas indicate that etching is a "cookbook" art.

X-Ray Data of Some Crystals

Table 6.5 is a list of Bragg angles of crystals of interest in solid state physics. The table is necessarily incomplete in two ways: not all crystals of possible interest are included, and not all the Bragg angles of any crystal are given. Crystals with small unit cells have few possible reflections for copper $K\alpha$ radiation. With diamond there are but five possible Bragg angles, but there are 35 orientations for these, not counting the back side of any plane as another Bragg reflection. If we use molybdenum $K\alpha$ ($\lambda = 0.7093$ Å) there are 34 possible Bragg angles with many many more orientations of these—such as 511, 151 ... 11$\bar{5}$ (12 of these) plus 333, $\bar{3}$33, $\bar{3}\bar{3}$3, and 3$\bar{3}\bar{3}$. All 16 orientations have a Bragg angle of 31°11′, again not counting the reverse side of any plane as another reflection.

Going to the other extreme, crystals with quite large unit cells have many possible reflections even for copper $K\alpha$ radiation. As an example, $BaNaNb_5O_{15}$ (orthorhombic, space group $cmm2$) has $a_0 \sim b_0 \sim 17.609$ Å. There are more than 410 possible $hk0$ reflections averaging one every 26 minutes of arc. Fortunately many are very weak, otherwise great confusion would reign. We can help matters by using a longer wavelength—for example, iron Kα (λ = 1.936042). This could have an $hk0$ reflection every 43 minutes of arc, but again many are very weak. We cannot go to much longer wavelengths because of air scatter unless we operate in a vacuum or a very light gas—not very practical. While we are on the subject of iron radiation, we should mention that crystals containing iron fluoresce badly in copper radiation, hence scatter radiation incoherently. Iron radiation does not do this with crystals containing iron.

TABLE 6.5   BRAGG ANGLES OF CRYSTALS OF INTEREST IN SOLID STATE PHYSICS, USING COPPER Kα X-RAYS

Silver arsenic sulfide, Ag_3AsS_3 (proustite), rhombohedral $R3c$, a_0 = 10.74 Å, c_0 = 8.658 Å

hkl	θ	I
11·0	8°15′	40
01·2	11°19′	5
21·1	13°41′	80
20·2	14°4′	70
30·0	14°23′	70
12·2	16°24′	100
11·3	17°38′	80
13·1	18°9′	90

$AgGaS_2$, $I\bar{4}2d$, a_0 = 5.74$_3$ Å, c_0 = 10.2$_6$ Å

hkl	θ
112	14°0′
103	15°12′
211	18°1′
220	22°18′
204	23°45′
301	24°10′
312	26°44′
116	29°16′
323	32°14′
400	32°27′

AgGaSe$_2$, Tetragonal, I$\bar{4}$2d

a$_o$ = 5.985, c$_o$ = 10.90

hk$l$	θ	I
112	13°19'	100
103	14°22'	10
211	17°14'	10
220	21°21'	75
204	22°29'	100
301	23°7'	10
312	25°32'	75
116	27°29'	25
323	30°30'	10
400	30°59'	25

AgI (Cubic)

Im3m, a$_o$ = 5.048

hk$l$	θ	I
110	12°28'	100
200	17°46'	60
211	21°57'	80
220	25°34'	10
310	28°41'	5
321	34°49'	10

AgI (Hexagonal)

P6$_3$mc

a$_o$ = 4.5922, c$_o$ = 7.510

hk$l$	θ	I
10.0	11°10'	65
00.2	11°50'	100
10.1	12°40'	40
10.2	16°23'	18
11.0	19°36'	85
10.3	21°19'	30
20.0	22°47'	8
11.2	23°9'	50
20.1	23°37'	6
20.2	26°0'	8

Ag$_3$SbS$_3$, Pyrargyrite

Silver Antimony Sulfide

Trigonal R3C

a$_o$ = 11.047, c$_o$ = 8.719

hk$l$	θ	I
01.2	11°12'	16
21.1	13°20'	55
20.2	13°50'	65
30.0	13°59'	55
12.2	16°4'	100
22.0	16°12'	10
11.3	17°26'	60
13.1	17°40'	60
31.2	19°52'	20

Aluminum Arsenide AlAs			Aluminum Antimonide AlSb		
Cubic F$\bar{4}$3M			Cubic F$\bar{4}$3M a_o = 6.1347		
a_o = 5.662 Å		MP			
hk$l$	θ	I	hk$l$	θ	I
111	13°38'	100	111	12°34'	100
200	15°47'	20	200	14°33'	34
220	22°38'	50	220	20°48'	74
311	26°49'	35	311	24°37'	31
222	28°7'	4	222	25°47'	5
400	32°58'	6	400	30°9'	5
331	36°22'	10	331	33°11'	15
420	37°28'	4	420	34°10'	10
422	41°48'	10	422	58°0'	21

Al$_2$O$_3$ (Corundum), Rhombohedral R$\bar{3}$c

Hardness 9 Mohs, MP2015°C

a_o = 4.75 Å c_o = 12.99 Å

	hk$l$	θ	I
	01.2	12°48'	74
	10.4	17°35'	92
X	11.0	18°55'	42
Z	00.6	20°50	< 1
	11.3	21°43'	100
	02.4	26°19'	43
	11.6	28°47'	81
	21.1	29°56'	3
	01.8	30°40	7
	21.4	33°19'	32
Y	03.0	34°11'	48

BaF$_2$ F.C. Cubic O$_h^5$ = Fm3m

MP 1280°C Slightly sol in Water

$$a_o = 6.184 \text{ Å}$$

hkℓ	θ	I
111	12°28'	100
X,Y,Z 200	14°26'	30
220	20°38'	79
311	24°24'	51
222	25°34'	3
400	29°53'	6
331	32°53'	13
420	33°51'	6
422	37°36'	14
333,511	40°20'	6
440	44°48'	2

(AlN)4H Hex P6$_3$mc

$$a_o = 3.114, \; c_o = 4.986, \; Z = 2$$

hkl	θ	I
100	16°36'	100
002	18°0'	60
101	18°57'	70
102	24°53'	20
110	29°39'	30
103	32°59'	20
200	34°50'	6
112	35°41'	18
201	36°17'	8

133

$Ba_2NaNb_5O_{15}$

Orthorhombic, cmm2

$a_o = b_o = 17.609, c_o = 3.9949$

hk$l$	θ	I	hk$l$	θ	I
240	$11°17'$	10	790	$29°55'$	10
001	$12°8'$	25	6,10,0	$30°40'$	40
150	$12°53'$	50	5,11,0	$31°54'$	50
221	$14°7'$	10	3,11,1	$32°46'$	10
131	$14°34'$	90	842(?)	$35°3'$	40
350	$14°47'$	90	5,11,1	$34°40'$	40
060	$15°13'$	10	392	$36°12'$	50
260	$16°4'$	10	0,10,2	$37°21'$	125
241	$16°41'$	100	3,13,0	$35°42'$	125
151	$17°51'$	40	592	$38°2'$	25
370	$19°28'$	10	133	$40°12'$	50
280	$21°9'$	25	243	$41°19'$	75
002	$22°51'$	90			
480	$23°2'$	40			
390	$24°31'$	40			
661	$25°15'$	10			
0,10,0	$25°56'$	100			
191	$26°30'$	50			
152	$28°25'$	25			
681	$29°2'$	75			
442	$29°12'$	50			
352	$29°27'$	100			
062	$29°42'$	25			
262	$30°13'$	40			
591	$29°48'$	10			

$Bi_{12}GeO_{20}$ (BGO) B. C. Cubic $I2_13$

MP 925°C Hardness Mohs 4.5

$a_o = 10.1455$

hkℓ	θ	I
110	6°10'	
200	8°44'	15
211	10°43'	
220	12°24'	17
310	13°53'	
222	15°15'	
321	16°30'	
400	17°41'	27
330	18°47'	12
422	21°50'	
440	25°26'	
600	27°6'	100
611	27°54'	
444	31°44'	
800	37°24'	13
811	38°5'	
660	40°6'	15

Bismuth Telluride BiTe

Cubic Fm3m, $a_0 = 6.47$

hkℓ	θ	I
111	11°54'	12
200	13°46'	100
220	19°41'	70
311	23°15'	6
222	24°21'	25
400	28°26'	12
331	31°16'	2
420	32°10'	30

$CaCO_3$, Calcite

Rhombohedral, $R\bar{3}C$

$$a_0 = 4.989 \text{ Å}, \quad c_0 = 17.062 \text{ Å}$$

hkℓ	θ	I
10.2	$11°32'$	12
10.4	$14°42'$	100
00.6	$15°43'$	3
11.0	$17°59'$	14
11.3	$19°42'$	18
20.2	$21°35'$	18
20.4	$23°33'$	5
10.8	$23°45'$	17
11.6	$24°15'$	17
21.1	$28°17'$	4
21.2	$28°42'$	8
10.10	$29°2'$	2
21.4	$30°20'$	5
20.8	$30°30'$	4
11.9	$30°41$	
21.5	$31°32'$	2
30.0	$32°20'$	5
00.12	$32°48'$	3

Calcium Fluoride CaF

Cubic FM3M

$$a_0 = 5.4626$$

hkℓ	θ	I
111	$14°8'$	94
220	$23°30'$	100
311	$27°53'$	35
400	$34°20'$	12
331	$37°56'$	10
422	$43°42'$	16

$$\text{Cd Cr}_2\text{S}_4, \text{ Cubic Fd3M}$$

$$a_0 = 10.207$$

hkℓ	θ	I
220	$12°19'$	55
311	$14°30'$	100
400	$17°34'$	18
331	$19°12'$	> 1
422	$21°42'$	30
511	$23°5'$	27
333	$23°5'$	27
440	$25°16'$	75
620	$28°31'$	14
533	$29°40'$	16
444	$31°31'$	6
551	$32°37$	< 1
711	$32°37$	< 1
642	$34°23'$	20
553	$35°26'$	12
731	$35°26'$	12
800	$37°8'$	12

CdGeAs$_2$, Tetragonal, I$\bar{4}$2m

$a_0 = 5.9427$, $c_0 = 11.2172$

hkℓ	θ	I
112	13°14'	100
103	14°5'	30
211	17°20'	30
220	21°30'	80
204	22°11'	100
301	23°15'	30
312	25°37'	80
116	26°48'	50
323	30°43'	30
400	31°14'	50
008	33°19'	30
332	34°32'	50
316	35°32'	50
424	39°54'	80

CdGeP$_2$, tetragonal, 1$\bar{4}$2m

$a_0 = 5.743$, $c_0 = 10.774$

hkℓ	θ
112	13°44'
220	22°18'
204	23°5'
312	26°35'
116	27°58'
224	28°22'
400	32°27'

β Cadmium Sulfide CdS

Cubic, $F\bar{4}3m$, $a_0 = 5.818$ Å

hkℓ	θ	I
111	$13^\circ 15'$	100
200	$15^\circ 21$	40
220	$21^\circ 59$	80
311	$26^\circ 3'$	60
222	$27^\circ 18'$	10
400	$31^\circ 59'$	20
331	$35^\circ 15'$	30
420	$36^\circ 18'$	10
422	$40^\circ 26'$	30

CdSe Hexagonal

$a_0 = 4.299$, $c_0 = 7.010$

MP>1350

	hkℓ	θ	I
Y	10.0	$11^\circ 56'$	100
Z	00.2	$12^\circ 42'$	70
	10.1	$13^\circ 33'$	75
	10.2	$17^\circ 34'$	35
X	11.0	$21^\circ 0'$	85
	10.3	$22^\circ 54'$	70
	20.0	$24^\circ 27'$	12
	11.2	$24^\circ 52'$	50
	20.1	$25^\circ 21$	12
	20.2	$27^\circ 56'$	8
	20.3	$31^\circ 57'$	20

CdSiP$_2$ Tetragonal I$\bar{4}$2d

$$a_0 = 5.678, \ c_0 = 10.423$$

hkℓ	θ	I
200	$15°45'$	20
004	$17°12'$	17
121	$18°12'$	13
220	$22°34'$	67
024	$23°39'$	100
301	$24°26'$	14
132	$26°59'$	46
116	$28°53'$	31
008	$36°15'$	49
332	$36°27'$	29
240	$37°21'$	13
136	$38°6'$	27

CdSnAs$_2$ Tetragonal I$\bar{4}$2d

$$a_0 = 6.092, \ c_0 = 11.922$$

hkℓ	θ	I
112	$12°45'$	100
220	$20°57'$	
204	$21°12'$	
312	$24°51'$	
116	$25°16'$	
400	$30°23'$	
208	$35°8'$	

CdSnP$_2$ Tetragonal I$\bar{4}$2d

a_o = 5.900, c_o = 11.513Å

hkℓ	θ	I
112	13°11'	100
200	15°8'	20
004	15°31'	8
220	21°40'	20
204	21°57'	55
312	25°43'	50
116	26°13'	18
224	27°8'	8
400	31°29'	6
008	32°22'	2
332	34°44'	8
316	35°10'	14
420	35°43'	10
404	35°56'	
208	36°33'	4
424	39°58'	16

Cr$_2$O$_3$ Hex R$\bar{3}$C

a_o = 4.954 c_o = 13.584

hkℓ	θ	I
012	12°16'	74
104	16°49'	100
110	18°7'	96
006	19°53'	12
113	20°46'	38
202	22°7'	9
024	25°8'	39
116	27°27'	90
112	29°14'	13
214	31°46'	35
300	32°35'	40
10.10	36°30'	20
220	38°28'	17
306	39°34'	7

CuCℓ (Nantokite)

Cubic F$\bar{4}$3m

a_o = 5.416

hkℓ	θ	I
111	14°16'	100
200	16°31'	8
220	23°43'	55
311	28°9'	30
400	34°40'	6
331	44°10'	8
511,333	47°49'	6
440	53°34'	2
531	57°17'	4
620	64°5'	4

Cadmium Telluride CdTe

Cubic F$\bar{4}$3m, a_o = 6.481Å

hkℓ	θ	I
111	11°53'	100
220	19°39'	60
311	23°13'	30
400	28°23'	6
331	30°35'	10
422	33°14'	10

CuCl (High Temp) Hexagonal $a_o = 3.91$, $c_o = 6.42$		
hk$l$	θ	I
10.0	$13°9'$	100
00.2	$13°53'$	80
10.1	$14°54'$	10
10.2	$19°18'$	30
11.0	$23°12'$	80
10.3	$25°12'$	70
11.2	$27°28'$	50
20.1	$28°4'$	20

CuFe$_2$O$_4$ Tetragonal, F4/ddm $a_o = 8.216$, $c_o = 8.709$		
hk$l$	θ	I
111	$9°10'$	30
202	$14°56'$	30
220	$15°23'$	15
113	$17°15'$	50
311	$18°1'$	100
222	$18°35'$	20
004	$20°43'$	25
400	$22°2'$	50
313	$23°27'$	10
331	$24°3'$	5
224	$26°14'$	15
422	$27°4'$	25
115	$27°30'$	25
333	$28°34'$	30
511	$29°5'$	50
404	$31°3'$	100
440	$32°2'$	50
531	$33°38'$	10
206	$34°15'$	10
620	$36°22'$	10

CdFe$_2$O$_4$ cubic Fd3m $a_o = 8.445$		
	hk$l$	θ
	111	$9°5'$
	220	$14°57'$
	311	$17°37'$
	222	$18°25'$
X,Y,Z	400	$21°24'$
	331	$23°26'$
	422	$26°32'$
	333,511	$28°17'$
	440	$31°4'$

CuGaS$_2$ Tetragonal

$$a_o = 5.36, \; c_o = 10.49$$

hkℓ	θ	I
112	$14°31'$	100
200	$16°42'$	2
004	$17°5'$	1
220	$23°59'$	60
204	$24°16'$	90
312	$28°32'$	90

α CuI, Cubic

$$a_o = 6.15$$

hkℓ	θ	I
111	$12°32'$	50
200	$14°30'$	30
220	$20°45'$	100
311	$24°33'$	90
222	$25°43'$	20
400	$30°4'$	20
331	$33°5'$	40
420	$34°4'$	30
422	$37°21'$	35
511,333	$40°36'$	25

CuInS$_2$ Tetragonal

$$a_o = 5.517, \; c_o = 11.05$$

hkℓ	θ
112	$13°59'$
004	$16°11'$
200	$16°13'$
204	$23°15'$
220	$23°16'$
116	$27°33'$
312	$27°35'$
224	$28°55'$

CuFeS$_2$ Chalcopyrite

Tetragonal I$\bar{4}$2d

hkℓ	θ	I
112	14°50'	100
020	17°6'	2
004	17°24'	2
220	24°34'	40
024	24°48'	80
132	29°14'	60
116	29°38'	10
033	29°39'	10
224	30°49'	5
040	36°1'	10
008	36°45'	5
332	39°54'	10
136	40°15'	15
143	40°16'	15
244	46°15'	60
228	46°46'	30

DyFeO$_3$ Orthorhombic

$$a_0 = 5.30, \; b_0 = 5.60, \; c_0 = 7.62$$

hkℓ	θ	I
110	$11°32'$	30
002	$11°40'$	30
111	$12°57'$	20
012	$14°9'$	2.5
102	$14°25'$	2.5
020	$15°58'$	20
112	$16°32'$	100
200	$16°54'$	20
021	$17°3'$	20
103	$19°39'$	5
211	$19°42'$	5
022	$19°58'$	10
202	$20°44'$	20
113	$21°18'$	20
122	$21°47'$	5
212	$22°19'$	2
220	$23°35'$	40
004	$23°51'$	40

ErFeO$_3$

Orthorhombic Pbnm

$$a_0 = 5.267, \ b_0 = 5.581, \ c_0 = 7.593$$

hkℓ	θ	I
110	11°36'	20
002	11°42'	10
111	13°1'	25
020	16°1'	20
112	16°36'	100
200	17°0'	6
021	17°6'	6
211	19°49'	18
022	20°2'	18
202	20°51'	16
113	21°24'	16
122	21°52'	2
212	22°27'	2
220	23°43'	25
004	23°56'	20

α Iron Oxide, Fe$_2$O$_3$ Hematite

Rhombohedral R$\bar{3}$c

$$a_0 = 5.0317 \ \overset{\circ}{A}, \ c_0 = 13.737$$

hkℓ	θ	I
01.2	12°5'	25
10.4	16°36'	100
11.0	17°50'	50
00.6	19°40'	2
11.3	20°27'	30
20.2	21°46'	2
02.4	24°45'	40
11.6	27°4'	60

$$\gamma \ Fe_2O_3$$

tetragonal, $a_0 = 8.338$, $c_0 = 25.01$

hk$l$	θ	I
101	$5°35'$	10
102	$6°23'$	20
110	$7°30'$	40
112	$8°18'$	10
113	$9°12'$	40
210	$11°55'$	60
213	$13°5'$	60
214	$13°55'$	15
205	$13°55'$	15
220	$15°9'$	90
300	$16°5'$	30
302	$16°30'$	20
310	$16°59'$	30
312	$17°22'$	20
313	$17°50'$	100

Fe_3O_4 - Magnetite

Cubic Fd3m, $a_0 = 8.396$

hk$l$	θ	I
111	$9°9'$	8
220	$15°2'$	30
311	$17°43'$	100
222	$18°32'$	8
400	$21°32'$	20
422	$26°43'$	10
511	$28°28'$	30
440	$31°16'$	40
531	$32°52'$	2
620	$35°28'$	4
533	$36°59'$	10
622	$37°29'$	4
444	$39°28'$	2
642	$43°21'$	4
731	$44°48'$	12

FeS_2 Marcasite

Orthorhombic Pnnm

$$a_0 = 4.445, \quad b_0 = 5.425, \quad c_0 = 3.386$$

hk$l$	θ	I
110	$12°57'$	40
020	$16°30'$	50
101	$16°37'$	50
111	$18°37'$	25
120	$19°26'$	25
210	$22°0'$	2
121	$23°46'$	30
211	$25°59'$	63
220	$26°37'$	4
002	$27°4'$	15
130	$27°23'$	10

Iron Sulfide FeS$_2$ Pyrite

Cubic Pa 3, a_0 = 5.417

hkℓ	θ	I
111	$14°15'$	36
200	$16°31'$	84
210	$18°32'$	66
211	$20°23'$	52
220	$23°43'$	40
311	$28°8'$	100
222	$29°31'$	14
230	$30°51'$	20
321	$32°9'$	24

GaAs Cubic Zinc Blende Structure F$\overline{4}$3m

Cleavage (110) a_0 = 5.64 $\overset{o}{A}$

hkℓ	θ	I
111	$13°41'$	100
220	$22°43'$	35
311	$26°56'$	35
X,Y,Z 400	$33°7'$	6
331	$36°32'$	8
422	$42°0'$	6
333,511	$45°12'$	4
440	$50°35'$	2

Ge (Germanium) Cubic, Dia. Structure Fdm3.

111 Cleavage MP 958°C

$a_0 = 5.65754$

hkℓ	θ	I
111	13°38'	100
220	22°39'	57
311	26°51'	39
X,Y,Z 400	33°0'	7
331	36°24'	10
422	41°50'	17
333,511	45°2'	7
440	50°22'	3

Germanium Telluride GeTe

Rhombohedral $a_0 = 4.17$, $c_0 = 10.66$

hkℓ	θ	I
Z 00.3	12°31'	20
10.1	13°1'	40
01.2	14°56'	100
10.4	21°3'	80
11.0	21°41'	80
01.5	24°48'	20
11.3	25°21'	20
02.1	25°38'	30
00.6	25°42'	80
02.4	31°1'	80

Mercury Sulfide, HgS, Cinnabar

Hexagonal

$$a_0 = 4.149 \text{ Å}, \; c_0 = 9.495 \text{ Å}$$

hkℓ	θ	I
10.0	$12°23'$	5
10.1	$13°15'$	100
00.3	$14°5'$	28
10.2	$15°36'$	94
10.3	$18°55'$	9
11.0	$21°48'$	26
11.1	$22°20'$	12
10.4	$22°53'$	29
11.2	$23°54'$	3
20.1	$25°52'$	21
11.3	$26°21'$	27
10.5	$27°19'$	25
00.6	$29°8'$	5

Mercury Telluride HgTe Coloradoite

Cubic F$\overline{4}$3m, $a_0 = 6.453$

hkℓ	θ	I
111	$11°56'$	100
200	$13°49'$	10
220	$19°44'$	90
311	$23°19'$	70
400	$28°31'$	20
331	$31°21$	30
422	$35°47'$	40
511,333	$38°20'$	30

151

HIO$_3$ Iodic Acid Orthorhombic P2$_1$2$_1$2$_1$

MP 110°C Water Soluble

a$_0$ = 5.53, b$_0$ = 5.92, c$_0$ = 7.75

	hkℓ	θ	I
	011	9°25'	15
	101	9°51'	31
	110	10°59'	100
Z	002	11°28'	9
	111	12°25'	38
	012	13°45'	43
	102	14°3'	53
Y	020	15°5'	7
	112	15°59'	31
X	200	16°11'	16
	021	16°10'	26
	201	17°12'	19
	120	17°10'	32
	210	17°54'	11
	121	18°9'	23

HoFeO$_3$ Orthorhombic Pbnm

$a_0 = 5.278$, $b_0 = 5.591$, $c_0 = 7.602$

hkℓ	θ	I
110	$11°35'$	80
002	$11°42'$	70
111	$13°0'$	80
020	$16°0'$	70
112	$16°34'$	100
200	$16°58'$	80
021	$17°4'$	70
211	$19°46'$	40
022	$20°0'$	60
202	$20°49'$	80
113	$21°22'$	70
122	$21°50'$	10
212	$22°24'$	10
220	$23°40'$	80
004	$23°55'$	60

InAs, Cubic F$\bar{4}$3m

$a_0 = 6.058$

hkℓ	θ	I
111	$12°43'$	100
200	$14°44'$	8
220	$21°5'$	60
311	$24°57'$	40
400	$30°34'$	10
331	$33°40'$	11
422	$38°32'$	16
33,511	$41°21'$	
440	$46°0'$	8

KH$_2$PO$_4$ Tetragonal I$\overline{4}$2d

a_o = 7.448 Å, c_o = 6.977 Å, MP 252.6°C

hk$l$	θ	I/I$_1$
101	8°42'	22
X,Y 200	11°56'	100
121	14°51'	12
112	15°21'	75
220	17°1'	23
202	17°37'	10
130	19°5'	3
301	19°14'	10
103	20°18'	5
231	22°53'	9
132	23°14'	37

LiF Cubic

Fm3m

a_o = 4.0270

hk$l$	θ	I
111	19°21'	95
200	22°30'	100
220	32°45'	48
311	39°23'	10
222	41°30'	11
400	49°55'	3
331	56°29'	4
420	58°48'	14
422	69°34'	13

α LiGaO$_2$ Trigonal R$\bar{3}$m

a_o = 2.911, c_o = 14.47

hkl	θ	I
003	$9°11'$	100
101	$18°4'$	40
006	$18°38'$	6
012	$18°53'$	16
014	$21°52'$	65
015	$23°54'$	14
009	$28°38'$	2
107	$28°48'$	12
018	$31°37'$	18
110	$31°57'$	18

LiIO$_3$ Hexagonal (soluble)

a_0 = 5.48, c_0 = 5.172

$$D_6^6 - C6_3 22$$

	hk$l$	θ	I
Y	01.0	$9°20'$	25
	10.1	$12°44'$	100
X	11.0	$16°20'$	30
Z	00.2	$17°20'$	8
	11.1	$18°33'$	2
	02.0	$18°57'$	8
	10.2	$19°50'$	10
	20.1	$20°56'$	18
	11.2	$24°11'$	25
	21.0	$25°26'$	4
	20.2	$26°9'$	4
	21.1	$27°2'$	20

155

LiNbO$_3$ (Lithium Niobate) Rhombohedral R3C
Cleavage (01.2)
Hardness 5 Mohs, M. P. 1253°C

$$a_0 = 5.15 \text{ Å}, \ c_0 = 13.864 \text{ Å}$$

hkℓ	θ	I
01.2	11°51'	100
10.4	16°21'	40
X 11.0	17°24'	20
Z 00.6	19°28'	4
11.3	20°2'	10
20.2	21°17'	10
02.4	24°15'	16
11.6	26°37'	20
12.2	28°3'	12
01.8	28°29'	6
21.4	30°32'	10
Y 03.0	31°12'	10
12.5	32°20'	
30.3	32°59'	
00.12	41°49'	
40.4	46°32'	
31.2	39°14'	
31.5	42°59'	

LiTaO$_3$ Rhombohedral R3C

$$a_0 = 5.159, \; c_0 = 13.76$$

hk$l$	θ	I
01.2	$11^\circ 52'$	70
10.4	$16^\circ 25'$	80
X 11.0	$17^\circ 22'$	70
20.2	$21^\circ 15'$	70
02.4	$24^\circ 17'$	70
11.6	$26^\circ 42'$	100
12.2	$28^\circ 1'$	80
01.8	$28^\circ 41'$	70
21.4	$30^\circ 32'$	70
Y 03.0	$31^\circ 9'$	70
20.8	$34^\circ 25'$	60
11.9	$35^\circ 51'$	70
22.0	$36^\circ 40'$	70
00.6	$19^\circ 36'$	

NaMnF$_3$ Orthorhombic, Pnma

$$a_O = 5.7485, \; b_O = 8.0045, \; c_O = 5.5509 \; \overset{o}{A}$$

	hk$l$	θ	I
	101	$11^\circ 7'$	95
	111	$12^\circ 27'$	4
X	200	$15^\circ 33'$	30
	121	$15^\circ 49'$	100
Z	002	$16^\circ 7'$	20
	201	$17^\circ 34'$	4
	102	$17^\circ 57'$	10
	211	$18^\circ 28'$	10
	031	$18^\circ 41'$	20
	112	$18^\circ 50'$	12
	022	$19^\circ 44'$	2
	131	$20^\circ 19'$	8
	221	$20^\circ 58'$	8
	122	$21^\circ 18'$	2
	202	$22^\circ 42'$	60

NaNiF$_3$, Cubic

$$a_O = 7.64$$

hk$l$	θ	I
200	$11^\circ 38'$	50
220	$16^\circ 34'$	100
311	$19^\circ 32'$	20
222	$20^\circ 27'$	20
400	$23^\circ 47'$	80
331	$26^\circ 4'$	20
420	$26^\circ 48'$	20
422	$29^\circ 36'$	60
440	$34^\circ 46'$	60

NaNiF$_3$ Orthorhombic Pbnm

a$_0$ = 5.36, b$_0$ = 5.53, c$_0$ = 7.71

hkℓ	θ	I
002	11°32'	50
110	11°33'	50
111	12°56'	25
020	16°11'	25
112	16°26'	100
200	16°42'	25
120	18°16'	25
210	18°38'	50
121	19°13'	25
103	19°25'	50
113	21°7'	50
122	21°49'	50
004	23°33'	75
221	24°22'	25
123	25°42'	25
130	26°13'	50

$NH_4H_2PO_4$ ADP Tetragonal $I\bar{4}2d$

$a_0 = 7.499$ Å, $c_0 = 7.553$ Å

	hkℓ	θ
	101	$8°19'$
	110	$8°21'$
Z	002	$11°46'$
X,Y	200	$11°51'$
	112	$14°30'$
	211	$14°33'$
	202	$16°50'$
	220	$16°53'$
	103	$18°50'$
	301	$18°56'$
	310	$18°57'$
	222	$20°48'$
	004	$24°5'$

NiS Hexagonal $P6_3mmc$

$a_0 = 3.44$, $c_0 = 5.35$

hk$l$	θ	I
10.0	$14°59'$	60
00.2	$16°44'$	10
10.1	$17°13'$	50
10.2	$22°46'$	100
11.0	$26°36'$	80
10.3	$30°14'$	40
20.0	$31°8'$	40
20.1	$32°28'$	40
00.4	$35°10'$	40
20.2	$36°17'$	70
10.4	$39°9'$	40
20.3	$42°22'$	20
21.0	$43°10'$	20
21.1	$44°21'$	60
11.4	$46°51'$	70

NiS Rhombohedral R3m

$$a_0 = 9.620 \quad c_0 = 3.149$$

hk$l$	θ	I
110	$9°13'$	60
101	$15°10'$	40
300	$16°6'$	100
021	$17°51'$	65
220	$18°41'$	12
211	$20°14'$	55
131	$24°25'$	95
410	$25°4'$	45
401	$26°19'$	40
321	$28°8'$	18
330	$28°43'$	35
012	$29°52'$	25
600	$33°'$	8
520	$35°16'$	4
312	$36°18'$	10

PbMoO$_4$ Tetragonal

$$I4_1/a \qquad a_0 = 5.435 \quad c_0 = 12.11$$

hk$l$	θ	I
101	$8°56'$	12
112	$13°44'$	100
004	$14°44'$	20
200	$16°28'$	25
211	$18°51'$	8
105	$20°23'$	6
213	$21°43'$	8

Si (Silicon) Cubic,
Dia. Structure Fdm3

MP 1420°C 111
   Cleavage

a_o = 5.4173 Å

	hk$l$	θ	I
	111	14°15'	100
	220	23°43'	60
	311	28°8'	35
X,Y,Z	400	34°40'	8
	331	38°18'	13
	422	44°9'	17
	333,511	47°38'	9
	440	53°33'	5

SiC (Hexagonal)

P6$_3$mc

a_o = 3.0763, c_o = 5.0480

hk$l$	θ	I
10.0	16°48'	60
00.2	17°46'	100
10.1	19°5'	80
10.2	24°52'	20
11.0	30°3'	---
10.3	32°47'	50
20.0	35°20'	10
11.2	35°54'	---
20.1	36°44'	20

β SiC (Cubic)

F43m a_o = 4.358

hk$l$	θ	I
111	17°50'	100
200	20°42'	20
220	30°0'	63
311	35°53'	50
222	37°45'	5
400	45°0'	6
331	50°24'	18
420	52°14'	6
422	59°59'	13
333,511	66°42'	10

SiO$_2$ (Quartz) Trigonal

P3$_1$2 and P3$_2$2

Transformation at 573°C

Hardness 7 Mohs

a$_o$ = 4.903 Å, c$_o$ = 5.393 Å

hkℓ	θ	I$_+$	I$_-$
Y 01.0	10°27'	24	24
01.1	13°21'	100	75
X 11.0	18°19'		
01.2	19°47'	28	85
11.1	20°11'		
02.0	21°16'	28	28
02.1	22°57'		
11.2	25°8'		
Z 00.3	25°22'	0.5	0.5
02.2	27°30'		
01.3	27°44'	45	2.7
11.3	32°6'		
03.0	32°58'		
02.3	34°9'	32	63
01.4	36°50'		
11.4	40°42'		
02.4	42°36'		
01.5	47°28'		
11.5	51°17'		
02.5	53°14'		
01.6	61°10'		

β tin Sn

Tetragonal D_{4h}^{19}

$a_o = 5.831$, $c_o = 3.174$

hkℓ	θ	I
200	$15°19'$	100
101	$16°2'$	90
220	$21°56'$	34
211	$22°29'$	74
301	$27°41'$	17
112	$31°20'$	23
400	$31°54'$	13
321	$32°19'$	20
420	$36°13'$	15
411	$36°36'$	15
312	$39°49'$	20

$SrTiO_3$ Cubic Pm3m, $a_o = 3.9051$

hkℓ	θ	I
100	$11°23'$	12
110	$16°12'$	100
111	$19°59'$	30
200	$23°14'$	50
210	$26°10'$	3
211	$28°54'$	40
220	$33°55'$	25
300	$36°17'$	1
310	$38°35'$	15
311	$40°52'$	5
222	$43°6'$	8
321	$47°34'$	16
400	$52°6'$	3
411	$56°49'$	10
331	$59°18'$	3
420	$61°54'$	10
332	$67°42'$	6
422	$75°5'$	9

$$Sr_xBa_{1-x}Nb_2O_5$$

Tetragonal P

For x = 3/4: a_o = 12.430, c_o = 3.913

hk$l$	θ	I
400	14°21'	75
330	15°15'	35
440	20°31'	6
350	21°11'	35
600	21°50'	14
260	23°4'	37
360	24°34'	50
601	24°53'	46
550	25°59'	61
170	25°59'	43
541	26°18'	45
142	28°0'	63
711	28°43'	54
800	29°43'	9
552	36°6'	48
770	37°50'	21
10,0,0	38°18'	48
880	44°31'	6
12,0,0	48°3'	12
004	51°57'	100
14,0,0	60°11'	10
0,14,1	62°50'	52

(+ Hundreds of others!)

For x = 0.6, a_o = 12.44, c_o = 3.942
For x = 0.5, a_o = 12.45, c_o = 3.949.
This is due to Ba ions being larger than
Sr ions. The Bragg angles are changed only
slightly by variations in proportions of Ba and Sr.

TeO_2, Tetragonal, $P4_2/mnm$

$a_0 = 4.810$, $c_0 = 7.613$ Sublimes at 450C

hk$l$	θ	I
101	$10°55'$	13
110	$13°5'$	85
111	$14°22'$	13
102	$14°57'$	100
112	$17°41'$	1
200	$18°41'$	20
201	$19°38'$	3
210	$21°59'$	3
211	$21°51'$	5
113	$22°15'$	
202	$22°16'$	
004	$23°53'$	11

TiO_2 (Rutile) Tetragonal $P4_2/mnm$

Decomposes at $1640°C$

Hardness $6-6.5$ Mohs

$a_0 = 4.5929$ Å, $c_0 = 2.9591$ Å

hkℓ	θ	I
110	$13°43'$	100
101	$18°2'$	50
X,Y 200	$19°36'$	8
111	$20°37'$	25
210	$22°2'$	10
211	$27°10'$	60
220	$28°19'$	20
Z 002	$31°22'$	7
310	$32°2'$	10
221	$32°45'$	2
301	$34°30'$	20
311	$36°13'$	2
202	$38°16'$	4
212	$39°55'$	2
321	$41°10'$	6
400	$42°8'$	4

Tourmaline (Dravite)

Rhombohedral R3m

$$a_0 = 15.931 \text{ Å}, \quad c_0 = 7.197 \text{ Å}$$

hkℓ	θ	I
10.1	$6°56'$	30
02.1	$8°54'$	25
30.0	$9°38'$	18
21.1	$10°31'$	65
22.0	$11°9'$	85
01.2	$12°47'$	60
13.1	$13°11'$	16
40.1	$14°20'$	6
41.0	$14°49'$	12
12.2	$15°4'$	85
32.1	$15°25'$	10
33.0	$16°52'$	> 1
31.2	$17°5'$	8
04.2	$18°1'$	100
24.1	$18°19'$	2

Y_3Al_2 (AlO_4) YAG (Garnet)

Cubic a_o = 12.008 Å Ia3d

	hkℓ	θ	I
	211	9°2'	35
	220	10°27'	10
	321	13°53'	20
X,Y,Z	400	14°52'	30
	422	18°19'	25
	431	19°6'	6
	521	20°34'	30
	440	21°17'	6
	611	23°19'	30
	631	25°47'	2
	444	26°23'	25
	640	27°33'	40

YFeO$_3$ Orthorhombic

$a_0 = 5.280$, $b_0 = 5.592$, $c_0 = 7.602$

hkℓ	θ	I
110	$11°34'$	108
002	$11°42'$	108
111	$13°0'$	70
020	$16°0'$	60
112	$16°34'$	100
200	$16°58'$	70
021	$17°4'$	40
211	$19°46'$	40
022	$20°0'$	20
202	$20°49'$	30
113	$21°22'$	40
122	$21°50'$	10
220	$23°40'$	70
004	$23°55'$	60
023	$24°13'$	40
221	$24°27'$	60

$Y_3Fe_2 (FeO_4)_3$ YIG Cubic, Garnet

$a_o = 12.376$ Å   Ia3d

	hkℓ	θ	I
	211	$8°46'$	35
	220	$10°8'$	10
	321	$13°28'$	20
X,Y,Z	400	$14°25'$	30
	422	$17°45'$	25
	431	$18°30'$	6
	521	$19°56'$	30
	440	$20°37'$	6
	611	$22°34'$	30
	631	$24°58'$	2
	444	$25°33'$	25
	640	$26°40'$	40

$ZnGeAs_2$ Tetragonal I$\bar{4}$2d

$a_o = 5.670, \ c_o = 11.153$

hkℓ	θ	I
112	$13°41'$	100
220	$22°36'$	50
204	$22°48'$	50
312	$26°49'$	50
116	$27°11'$	25
400	$32°55'$	25
008	$33°32'$	10
332	$36°21'$	10
316	$36°39'$	25
424	$41°52'$	50

172

ZnGeP$_2$

Tetragonal I$\bar{4}$2d

$a_o = 5.465$, $c_o = 10.771$

hkℓ	θ	I
112	14°13'	63
020	16°23'	20
004	16°38'	21
220	23°30'	92
024	23°42'	100
132	27°56'	50
116	28°16'	52
224	29°25'	14
040	34°21'	65
008	34°56'	70
136	38°15'	34
240	39°7'	9
044	39°15'	9

ZnO Hex P6$_2$mc

$a_o = 3.249$, $c_0 = 5.205$

hkl	θ	I
100	15°53'	71
002	17°13'	56
101	18°8'	100
102	23°47'	29
110	28°18'	40
103	31°26'	35
200	33°12'	6
112	33°59'	28
201	34°33'	14
004	36°18'	3
202	38°29'	5
104	40°42'	3

ZnS (Sphalerite)

Cubic F$\bar{4}$3m

$a_0 = 5.3918$

hk$l$	θ	I
111	$14°20'$	100
200	$16°36'$	10
220	$23°50'$	51
311	$28°17'$	30
222	$29°40'$	2
400	$34°51'$	6
331	$38°31'$	9
420	$39°43'$	2
422	$44°25'$	9
511	$47°56'$	5
440	$53°55'$	3
531	$57°41'$	5
620	$64°38'$	3
533	$69°31'$	2

ZnS, Wurtzite

Hexagonal, P6$_3$mc

$a_0 = 3.820$, $c_0 = 6.260$

hk$l$	θ	I
100	$13°28'$	95
002	$14°15'$	50
101	$15°16'$	100
102	$19°48'$	40
110	$23°47'$	65
103	$25°53'$	60
200	$27°45'$	13
112	$28°12'$	45
201	$28°48'$	20

ZnSe Hex

$$a_0 = 3.996 \quad c_0 = 6.53$$

hkℓ	θ	I
100	$12°52'$	100
002	$13°39'$	90
101	$14°35'$	70
102	$18°56'$	60
110	$22°41'$	100
103	$24°43'$	80
112	$26°54'$	70
202	$30°17'$	10
203	$34°41'$	60
120	$36°7'$	30
105	$39°7'$	40
300	$41°56'$	40
213	$43°26'$	60
302	$45°8'$	40
205	$47°42'$	20
220	$50°30'$	30
116	$53°46'$	50
215	$56°32'$	20
313	$61°23'$	20
401	$63°0'$	20

ZnSiAs$_2$ Tetragonal, I$\bar{4}$2d

$a_o = 5.60$, $c_o = 10.88$

hk$l$	θ	I
011	8°54'	1
112	13°55'	63
013	14°39'	3
020	15°18'	3
004	16°27'	3
121	18°24'	3
015	22°19'	2
220	22°54'	100
024	23°15'	98
031	24°45'	4
132	27°13'	45
116	27°51'	40
224	28°46'	3

ZnSiP$_2$

Tetragonal, I$\bar{4}$2d

$a_o = 5.407$, $c_o = 10.451$

hk$l$	θ	I
112	14°27'	100
200	16°33'	6
220	23°46'	96
204	24°12'	76
132	28°18'	40
116	29°5'	33
400	34°44'	48
008	36°8'	46
332	38°28'	25
424	44°35'	32
228	45°34'	35
440	53°42'	24
408	55°5'	25

ZnSnAs$_2$ Tetragonal

$$a_o = 5.852, \quad c_o = 11.705$$

hk$l$	θ	I
101	8°28'	20
112	13°11'	100
103	13°44'	20
004	15°16'	5
200	15°16'	5
211	17°33'	20
105	20°45'	10
213	20°45'	10
204	21°51'	50
220	21°51'	50

ZnTe Cubic

$$F\bar{4}3m, \quad a_o = 6.1026$$

hk$l$	θ	I
111	12°38'	100
200	14°37'	10
220	20°55'	80
311	24°45'	35
222	25°56'	4
400	30°19'	8
331	33°23'	14
420	34°22'	4
422	38°12'	10
511	40°59'	8

PbTa$_2$O$_6$ Hex (Rhomb) R3m

a$_0$ = 10.55 , c$_0$ = 11.35

hkℓ	θ	I
10.1	6°13'	2
11.0	8°24'	2
01.2	9°12'	2
02.1	10°28'	6
00.3	11°45'	6
11.3	14°31'	100
03.0	14°39'	100
12.2	15°8'	2
10.4	16°31'	2
22.0	16°59'	2
03.3	18°57'	4
40.1	20°7'	2
22.3	20°51'	25
04.2	21°19'	2

PbZrO$_3$ Ortho   Pba$_2$

a$_0$ = 5.88 , b$_0$ = 11.75 , c$_0$ = 8.21

hkℓ	θ	I
110	8°25'	4
120	10°41'	14
002	10°49'	8
130	13°40'	2
112	13°46'	6
040	15°12'	55
122	15°18'	100
210	15°40'	6
132	17°34'	2
042	18°49'	18
230	19°7'	2
212	19°12'	2
240	21°45'	30
004	22°3'	16

$Ba_2NaNb_5O_{15}$

The $Ba_2NaNb_5O_{15}$ crystal is tetragonal above 300°C; if cooled below this temperature it becomes a complexly twinned array of orthorhombic domains. However if one of the "a" axes is subject to a uniform compressive stress during cooling, a single domain results. The stressed "a" axis becomes the orthorhombic "b" axis while the "c" axis remains as the "c" axis. The literature gives $a_0 = b_0 = 17.609$ Å for the orthorhombic crystal. Since the crystal is biaxial, it cannot be tetragonal. Hence although X-ray tests cannot distinguish the "a" axis from the "b" axis, optical tests can accomplish this: the two optic axes are perpendicular to the "a" axis but make angles of about 6 degrees with the "b" axis. Some investigators have described this crystal with axes 45° from the ones used here (i.e., class $mm2$), but this unit cell does not have the symmetry of the crystal. Between crossed polarizing filters an orthorhombic crystal should have the "a" and "b" axes as extinction directions. The $cmm2$ cell has this, the $mm2$ cell has a 45° extinction angle. With the twinned $cmm2$ cell, twin boundaries are 110 planes. With the $mm2$ cell, twin boundaries are 100 planes.

7

Crystal Sawing

SAWING

Sawing is commonly the first operation in reducing a piece to required dimensions. The method of sawing depends on the material.

Muck Saws

In the manufacture of quartz oscillator plates muck saws were used before diamond saws proved their practicality. A muck saw is a revolving metal disk, usually steel, that dips into a suspension of abrasive in fluid. The piece to be cut is pressed against the edge of the disk and a slot is abraded into the piece. The disk is rotated without throwing off the abrasive that clings to the disk. This limitation due to centrifugal force reduces the cutting speed to a fraction of that obtainable with diamond saws. Figure 7.1 illustrates such a saw built from a belt-driven grinder. The mud pan does not appear. The end view shows a rotatable arm that allows one to make angle cuts. The arm turns about the shaft S to allow the work to press against the periphery of the saw. For unattended

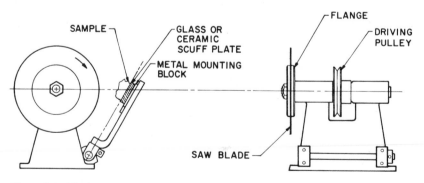

Figure 7.1   Muck saw.

operation, weights are used to force the work gently against the saw. These weights can be hung on a second arm that attaches to the arm shaft at its right end, as seen in the frontal view. Inasmuch as vibrations can build up in such a system, it is well to have a dash pot with heavy oil to dampen such vibrations. If the pressure is hand applied, the hand is a sufficient damper.

The muck saw is slow and messy but it can cut almost anything except diamond; the fluid needs to be chosen to suit the material. For example, water would not be used on a crystal that is highly soluble in water. However water that has dissolved all this crystal that it can dissolve (i.e., a saturated solution) works well. The saturation can be supplied by dissolving small scraps of the material. Such a saturated solution is generally superior to an oil for this cutting job in that it causes less cracks to form—probably because most soluble materials are more soluble at higher temperatures. In the saw kerf, heat is generated, and the fluid becomes less than saturated; hence solution occurs and frees the saw. With an oil in which the crystal is insoluble, no freeing action accompanies the increase in temperature.

Another disadvantage of muck saws is that a wedge-shaped kerf is commonly produced. The saw blade is generally made from sheet steel and is of uniform thickness to begin with. Abrasion thins down the edge, causing a cut wider near the top than near the bottom. Also as the blade plunges in deeply, abrasive continually flows over the mouth of the cut, further widening it. For this reason precise orientation is not held, and each piece may have to be reoriented if the tolerances are narrow. The abrasive readily works its way into the nearest main bearing, rapidly destroying the saw.

Cutoff Saws

An abrasive cutoff wheel is extremely good for soft materials, acting effectively as a very thin grind stone. The abrasive grains can break free from the bond, leaving fresh sharp grains exposed. The cutting is hence very free if used with a saturated solution. We have cut thousands of slices of ammonium dihydrogen phosphate with such a saw. Some of the crystals were 6 in. square in cross section. We used as fluid a mixture of propylene glycol and water, which was then saturated with the crystalline material. These saws may be run on a machine like that of Fig. 7.1, but the speed should be increased. The saw should be accurately centered or trimmed with a diamond point, to avoid all pounding. The saws may also be run on a wet surface grinder, but such machines are much more expensive than that of Fig. 7.1. A good stream of fluid should impinge on both sides of the saw, as explained below.

To aid in choosing cutoff wheels we now discuss abrasive wheel nomenclature. There are fundamentally five areas specified in ordering a grinding wheel: abrasive type, grain size, grade, structure, and bond type. By the 1958 American Standards Convention, A designates an aluminum oxide wheel, C a silicon carbide wheel. The manufacturer may prefix this by symbols identifying his particular abrasive. The grain size appears next in terms of the numerical series given previously. The "grade" specifies the strength of the bond. Letters are used as follows: E, F, G are "very soft"; H, I, J, K, are "soft"; L, M, N, O are "medium"; P, Q, R, S are "hard"; and T, U, W, Z are "very hard." We remark that J or K wheels are recommended for grinding agate, the type being silicon carbide (C). Structure specifies the porosity of the aggregate. The designations are numbers running from 1 to 15 or higher, with the larger numbers used for the more open structures. The structure designation is often omitted. Bond types are V (vitrified), S (silicate), E (shellac or elastic), R (rubber), RF (rubber reinforced), B (resinoid), BF (resinoid reinforced), and O [(magnesium) oxychloride]. Sometimes the manufacturer adds another number for his records. An example of a cutoff wheel for crystal slicing is C120KR, a silicon carbide, size 120 grit, of soft designation, rubber bonded. Hard wheels are generally used for soft materials and soft wheels for hard materials, since the hard materials require more frequent replacement of cutting grains.

Cutoff wheels are frequently shellac bonded, but rubber is also good. Elasticity is needed in such thin wheels. Silicon carbide wheels are used for cutting tile, ceramics, and slate, but diamond wheels are faster for these materials.

Saw Flanges

Any saw used on crystals should be supported by flanges that extend as near to the edge as is practical. Abrasive cutoff wheels and diamond saws often are used with slinger flanges, illustrated in Fig. 7.2. Fluid can be sprayed into the external channel; centrifugal force throws it out through the grooves that are cut from the opposite side, deeply enough to break through into the channel. Generally such flanges are relieved slightly near the center of the contact face so that the pressure is greatest near the edge of the flange.

Diamond Saws

The simplest diamond saw is a soft metal disk into the edge of which diamond powder has been pressed. This saw performs well on a machine such as that of Fig. 7.1 if the mud pan is replaced with a clean water pan and water is sprayed against the blade. The saw should run at perhaps

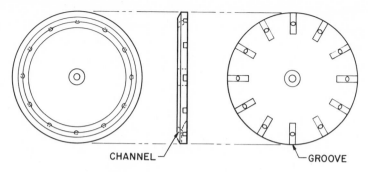

CHANNEL   GROOVE

Figure 7.2   Slinger flange.

1000 surface feet per minute and will require splash guards to keep the water from being broadcast. There are two common methods of charging the disk. In both methods diamond dust of about $40\,\mu$ or finer is suspended in olive oil and a little of this suspension is rubbed into the periphery of the wheel. Some operators merely cut a piece of agate free hand to force the diamonds into the metal, others use a hard steel roller to press the diamonds in. The roller makes the edge of the saw thicker than the middle, a real advantage.

The next degree of sophistication is to notch the periphery of the blade by chopping it with a sharp heavy knife. The diamond in olive oil is rubbed into the periphery as before, and the roller closes the cuts, trapping some diamond rather deeply. Such blades, made for the sawing of tile, are available commercially.

A superior blade is produced by the technique of powder metallurgy. Diamond and bronze powders are compressed around the edge of a bronze disk with quite high pressure. Baking sinters the metal giving a fairly strong bond. These saws are more expensive but give superior performance. They should be used only on a fairly rigid surface grinder type of machine, since the diamond and powder saw edge is somewhat brittle.

Saw blades used on a machine like that in Fig. 7.1 plunge directly into the work while the surface grinder type of saw makes a series of passes over the work, cutting a little deeper on each pass. The first type of cutting is called plunge cutting, the second is called traverse cutting.

When a very thin blade is essential, an internal saw can be considered. Here the blade can be soldered to the edge of a cup (see Fig. 7.3). Because of differential expansion, cooling after soldering puts tension in the saw, making it less likely to bend in cutting. Such a saw is limited to the cutting of small pieces by plunge cutting, traverse cutting being impossible.

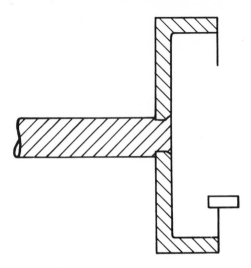

Figure 7.3   Internal saw.

The Traverse Saw

The traverse saw is a surface grinder adapted to wet sawing. A schematic diagram is given in Fig. 7.4. The machine is essentially an L frame (L, Fig. 7.4), which carries a block that can slide up and down in a recess in the vertical leg of the L. This block carries the bearing for the saw shaft. The shaft is turned by an electric motor (not shown) behind the L frame. The block is raised and lowered by means of a lead screw (not shown). The horizontal leg of the L frame has a male vee to guide a second block B2 that slides in and out, as indicated by the arrows on the side of B2. The block is slid in and out relative to the L frame by means of a second lead screw (not shown). The block B2 also has a male vee that guides the bed B as it slides back and forth, forced by means of a rack and pinion (not shown). Commonly a magnetic chuck is fastened to bed B and the mounting block on which the crystal is cemented is placed on this chuck. The barrel holder (Figs. 4.15 and 4.16) can be clamped in a large vee block on the magnetic chuck. In this way plates of required orientation can be sawed directly from the mother crystal. A watertight cover keeps the water that is sprayed onto the saw from being broadcast and from getting into the ways of the machine. The in-and-out motion of block B2 and the vertical adjustment of the saw can be set to a fraction of a mil. This gives fine control of slice thickness during sawing and also when the machine is used with a wheel in place of the saw (i.e., as a surface grinder). If one has much precise cutting to do, the traverse saw is a good investment.

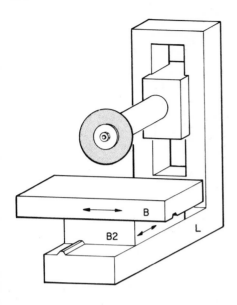

Figure 7.4 Traverse saw.

String Saws

String saws have some advantages over abrasive saws. For example, they are very gentle in action. A string saw can leave a crystal at a very sharp angle, producing a wedge of a few degrees angle. The apex will be razor sharp. If the same cut were attempted with an abrasive saw, the wedge would break off before the cut was completed. An acid string saw can cut metals that would be difficult to cut by abrasion. The surface damage, always present when abrasive saws are used, is absent with string saws. Also, the forces involved are very small. Crystals can be fastened down by pressing them against one or more small clay balls. An abrasive saw needs better securement—generally molten wax is applied to the hot crystal and supporting block assembly.

A very simple string saw is schematized in Fig. 7.5. An endless string passes over three small pulleys, A, B, C. The crystal to be cut presses against the under side of the string. Pulley C dips in water, moistening the traveling string. As the string travels, it dissolves its way through the crystal. The crystal is supported by the arm D, which is pivoted about an axis at P. A weight W causes the crystal to press against the string. The string travels 50 to 100 feet per minute. A sponge is commonly applied to the string between pulleys B and C to keep the string from becoming too wet. To check for proper degree of wetness, pinch the moving string lightly between the fingers for a moment and inspect the fingers for moisture; there should be only a trace of moisture where the string

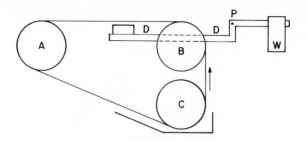

Figure 7.5 Simple string saw.

touched the fingers. If the string carries excessive moisture, the cut will be too wide and inaccurate.

There are many variations on this simple string saw. For use with acid the pulleys are made of plastic; the string is also made of plastic that will tolerate the acid. The last requirement is avoided in the saw in Fig. 7.6. Cotton, linen, or any appropriate thread is pulled from spool A. In passing under pulley B, the thread dips in acid, passes over pulley C, over the crystal being cut, and over pulley D; then it is taken up on spool E. The thread travels much more slowly than the string of Fig. 7.5. The string is thrown away when spool E gets full. A brake is applied to spool A, and spool E is driven by a small-geared motor. In this way the thread is sufficiently tight to ensure a true cut.

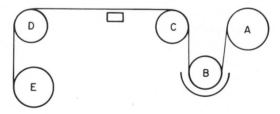

Figure 7.6 Acid string saw using disposable thread.

Two other designs that eliminate the requirement for endless strings appear in Figs. 7.7 and 7.8. In Fig. 7.7 the ends of the string are tied into a bead. As the bead approaches the crystal, it strikes lifter L, a bent plate with a slot in it. The string passes through the slot readily but the bead will not. The string is lifted from the cut while the bead passes over the lifter, then the string is dropped back into the cut. Hence we have a pseudo-

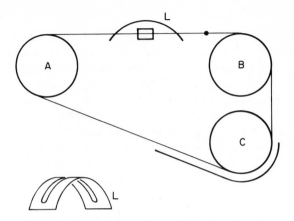

Figure 7.7   Bead lifter, used to avoid splicing the string.

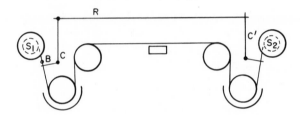

Figure 7.8   Reversing string saw.

endless belt. The groove in the pulleys must be deep enough to prevent the bead from throwing the string out of the grooves. The second scheme is a little less simple. In Fig. 7.8 we see two spools S_1 and S_2. One end of a long string is fastened to spool S_1, the other to spool S_2. A few inches from each tied end of the string is a bead. As one spool winds up the string, the other spool lets the string out. Just before the string is completely unwound from the first spool, the bead B comes out and presses against a forked bell crank lever C, which pushes a connecting rod R. This action declutches the driving spool and engages a clutch, which then turns the spool that has just unwound. Hence this spool rewinds until the bead on the other end pushes a second bell crank C' at that end, pushing the connecting rod R back to its first position, hence re-reversing the action. The string runs from spool to spool over and over again. Two solvent supplies are necessary.

Wire Saws

Wire saws are a special case of string saws. Commonly a fine wire is used and abrasive instead of solvent does the cutting. To reduce the loss of material of the saw kerf, the wire must be of small diameter. To cut straight, it should be stretched tightly. This suggests that the wire must be very strong. Tungsten is often used as a wire fulfilling the requirements of having great strength with a small diameter. The ends of tungsten wire can be beveled and brazed together to get an endless wire. Figure 7.9 is a schematic representation of a commercially available wire saw.

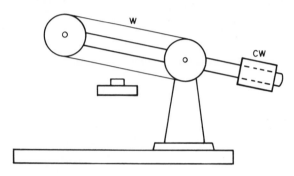

Figure 7.9 Wire saw.

Wire saws are not limited to small workpieces. Slabs of slate, marble, and novaculite, can be sawed from the natural site; hills are literally sawed in two by a stranded cable that runs over pulleys and is fed with sand and water. The strands act like grids cut into a lap and give entry points for the abrasive.

A String Gang Saw

Several novel features are involved in string gang saws, once used in making telephone channel filters. The material was ethylene diamine tartrate (Fig. 7.10*a*), and the plates had to be cut from the crystal at complicated angles—to get a zero temperature coefficient of frequency. To save material by cutting the final slabs directly from the crystal, a "comb" (Fig. 7.10*b*) was used to support the crystal. The strings of the saw passed between the teeth of the comb, the teeth being wide plates, each with a notch cut into it so that the crystal could be merely set in the notches and would then be correctly oriented. This synthetic crystal, called EDT, grew with natural faces so nearly perfect, that orientation could be established with quite good accuracy in this way. A little clay

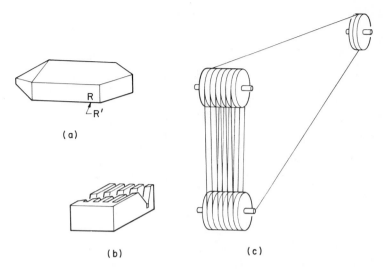

Figure 7.10   String gang saw.

was smeared over the two reference faces to hold the crystal in place. The clay must be water bound, not grease bound. The modeling clays (plasticene, plastilene, etc.) are grease bound, and grease is bad for the water-carrying string. The faces marked R and R' in Fig. 7.10a were the reference faces to be contacted to the comb groove walls. Since the crystals grow with very sharp edges, the bottom of the grooves were relieved as shown.

The saw was made from a pair of multigroove pulleys over which the string passed (Fig. 7.10c), one string serving for all cuts. This made one string splice do for all strings. The rear slack takeup pulley or idler allowed for adjusting the string tension.

The upper left-hand pulley was driven by a small-geared motor at a speed that gave 50 to a 100 feet per minute string speed. The comb was clamped to a small carriage sliding on a track and was propelled by a second small-geared motor. If the cutting portion of the strings was deflected by an arbitrary small amount, the motor was interrupted by placing a sapphire bar B back of the strings (see Fig. 7.11). When pushed back the bar pressed a microswitch ms that stopped the motor. This arrangement proved advantageous because it is difficult to estimate the drive speed appropriate to a given-size crystal, hence a variable speed drive gearbox is impractical. If the strings do not cut as fast as the crystal advances, the strings are increasingly deflected and can no longer cut true. The deflection-controlled drive was quite satisfactory.

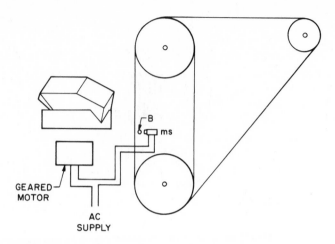

Figure 7.11 Crystal advance controlled by string deflection.

MAKING ENDLESS STRINGS

To make the endless strings we built a special winder (Fig. 7.12). Three
pulleys D, E, F determine the length of the belt; F was adjustable. A fine
thread, generally nylon, was wound on a bobbin B, and the bobbin was
mounted on a fourth pulley PG. This pulley had a central hole through
which the belt passed, but the pulley, shaft, and bearing were slotted so
that a finished belt could be removed. Pulley D was driven by motor M by
means of worm W and worm gear WG. Motor M also carried a pulley Pg.

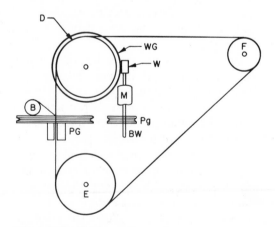

Figure 7.12 Making endless strings.

To start a belt, the slot in the pulley was lined up with the slot in the bearing and thread was pulled from the bobbin, passed around pulleys E, F, and D, and tied to the thread near the bobbin. In the pilot model a piece of grocer's twine was then tied around pulleys PG and Pg to drive pulley PG. The motor was turned on and the bobbin wound thread on the belt until the belt was big enough. A few inches more of thread was pulled from the bobbin and the thread cut. This loose end was sewed into the belt with a needle, and the belt was marked with ink to show the direction the string should travel in the saw. A dash-dot was adopted, the dot being imagined to be the feather end of an arrow, the dash the head. It was thought that it would be better to have this end drag than have it snub and pile up into a knot. With the marking completed, the grocer's twine was cut and the belt was removed from the winding machine. In subsequent industrial models the pulleys PG and Pg were replaced by spur gears, the gear PG being slotted between two teeth clear to the bearing center to allow for belt removal. With this model no twine belt was needed.

Nylon was a very satisfactory material for endless belts because of its stretchability. The stem BW served for rewinding bobbins.

HOLDING FIXTURES FOR ANGLE CUTTING

In a simple cross slide used on a saw (Fig. 7.13), the metal work-holding plate is clamped in the gap G with set screws SS. The work is advanced between cuts by turning the lead screw attached to the graduated drum D on which thousandths of inches are read directly.

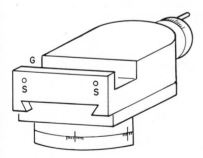

Figure 7.13 Crystal holding fixture, for use in a string saw.

More complicated devices are available whereby a crystal may be turned about two or three different axes to achieve special angles of cut. Such a device appears in Fig. 7.14. Special orienting fixtures have been described in connection with X-ray orienting.

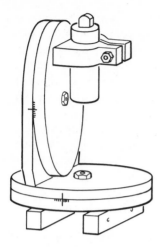

Figure 7.14   Crystal holding fixture for sawing.

SAWING LARGE, THERMALLY SENSITIVE CRYSTALS

Soft crystals often have large coefficients of thermal expansion coupled with low strength and poor thermal conductivity. Also they are sometimes heavily stressed because of growing conditions. Such crystals easily crack if they are heated to facilitate connecting them to a mounting plate prior to sawing. To avoid this hazard the crystals can be mounted in dental plaster and then sawed with a gang-muck saw. To allow this to be done with precise orientation, a convenient device (Fig. 7.15) consists of a handle piece and two links that supply three mutually perpendicular axes. The upper end of the handle piece simply allows the device to be carried safely from the X-ray machine to a transfer fixture after X-ray orientation is accomplished. The lower end of the handle is a reference cylindrical surface that fits into the vees of the X-ray goniometer and later into the vees of the transfer fixture. The axis of the handle is marked as the axis H1 in the illustration. With the handle in the goniometer vees, this axis is horizontal. Attached to the lower end of the handle piece is the link L1. This link provides an axis for the second link L2, and this axis is labeled H2 in Fig. 7.15. The second link carries a third axis, labeled V in the illustration. The third axis carries a saddle piece SP in which the crystal boule is secured by means of masking tape—after a rough orientation made by noting rudimentary faces on the boule sides. The saddle piece, hence the length of the crystal, make an angle of 45° with the axis V.

We describe the X-ray orientation process as starting with axes H1 and H2 horizontal and axis V vertical, the handle piece being in the vees of the X-ray goniometer. An X-ray beam XR strikes the outer surface of the

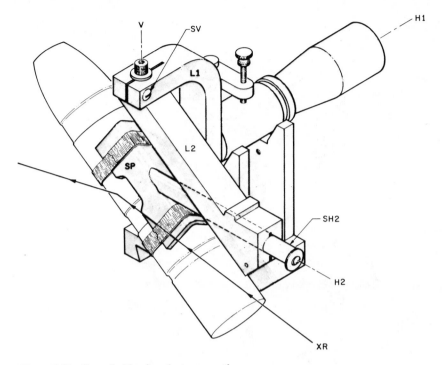

Figure 7.15 X-ray holder for plaster mounting.

crystal, and the shaft of axis V is turned until the goniometer arm reads the proper Bragg angle at maximum X-ray counting rate. The shaft is then locked by means of locking screw SV. The handle piece is now turned 90° in the vees so that axis H2 is now vertical and axis V is horizontal. In this position link L2 is adjusted to give the maximum counting rate at the proper Bragg angle setting, whereupon screw SH2 is tightened to lock the axis. It is well to check both readings at 180° to the position just described.

The oriented crystal in its adjusted cradle fixture is now placed in the transfer fixture (Fig. 7.16), with the crystal horizontal as shown. A mounting plate that will be transferred to the saw is clamped onto the riser platform RP. Masking tape is placed around the edges of the mounting plate, and a freshly mixed batch of dental plaster is put into the resulting enclosures. After loosening locking screw LS, the mounting plate and its supporting platform are lifted up until the crystal is partially buried in plaster; screw LS is then tightened so that the crystal remains buried. In about 15 minutes the plaster will be so hard that the cradle fixture can be safely removed, by cutting the masking tape near the saddle. The

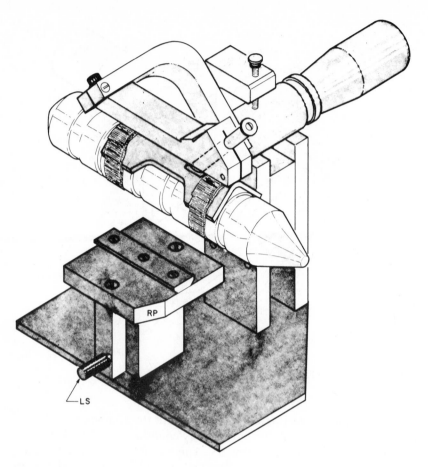

Figure 7.16   Transfer fixture.

mounting plate is now transformed to the saw (Fig. 7.17), and the cutting is started.

The saw is essentially a set of thin metal strips stretched in a metal gang frame, supported by four bars dropping down from two cross blocks. These cross blocks are carried by two slide rods that slide in slide bearings as reciprocated by the geared motor. The individual blades are stretched tight by means of wedges W, driven in firmly. Gang frames can be replaced with other frames with different saw separations. This enables the operator to cut the edge cuts on a "cant" then turn the "cant" 90°, reorient with X-rays, change gang frames, and slice the "cant" into thin slices. The word "cant" used here is from the lumber industry. A log

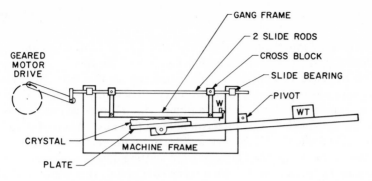

Figure 7.17 Crystal on mounting plate transferred from transfer fixture (Fig. 7.16) to saw, ready for cutting.

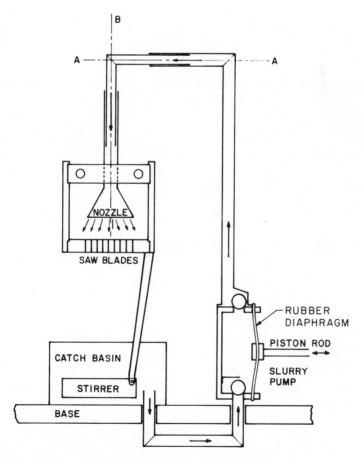

Figure 7.18 Apparatus for the automatic application of abrasive.

is often sliced into "cants" of thickness of 4, 6, or 12 in. The cants are run
through a gang saw that slices them into 1 in. boards.

The abrasive is applied automatically, as in Fig. 7.18. A drain pan
carrying a slurry of abrasives in water catches slurry falling from the saw
blades. The slurry drains into the pump which forces it up to be sprayed
over the saw blades. A catch basin holds a thin suspension of abrasive in
water, and a stirrer driven back and forth by the reciprocating motion of
the saw frame keeps the abrasive from settling during saw operation. The
slurry is pumped from the catch basin to a nozzle that squirts the slurry
over the saw blades from whence it drains back into the catch basin. The
pump is a rubber diaphragm pump with two ball check valves. The piston
rod is driven in and out by means of a bell crank, which in turn is driven
by an eccentric disk on the shaft that drives the saw frame. Consequently
the saw blades receive one squirt per stroke. The nozzle is a flattened
cone and can be positioned by swinging about the axis A–A and
foreshortened by turning it about the axis B. Because the lower ball valve
is lower than the level of the slurry in the catch basin, the pump is always
"primed."

The device of Fig. 7.15 is used for large, thermally sensitive crystals.
Figures 7.19 and 7.20 show a very small version suitable for transferring
from the X-ray goniometer to a wire saw for sawing parallel to a required
atomic plane. The crystal "xtl" can be stuck to the "tongue" by means of
cheese wax. The tongue goes into the receiver (Fig. 7.19) and is locked
(lock not shown). The axle A_2 goes into the bearing link, and the bearing
link goes onto axle A_1 of the base link. The axles can be locked by means
of screws that bend the split ends of the bearing link to clamp the axles.

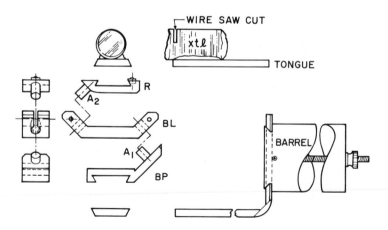

Figure 7.19   Small machine for transferring a crystal from X-ray goniometer to wire saw.

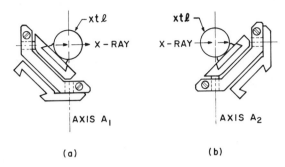

(a)   (b)

Figure 7.20   Positions assumed in adjusting orientation of a machine like that of Fig. 7.19.

The base link slides onto a tongue that bends at right angles to be clamped to a barrel that fits into the X-ray goniometer.

   In adjusting the orientation, the array is placed in the goniometer vees in the position illustrated in Fig. 7.20a ; axis A_1 is unlocked, and the upper assembly is swung about this axis to a position where the counting rate meter reads a maximum for this position and also for a position 180° around the barrel axis from Fig. 7.20a. Axis A_1 is now locked and the

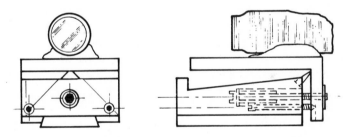

Figure 7.21   Modification of barrel holder (Fig. 4.15) for use in wire saw adjusted as in Fig. 7.20.

assembly is placed in the position shown in Fig. 7.20b. The orientation adjustment is then made as before and axle A_2 is locked. The assembly (without the barrel and its tongue) are now transferred to the wire saw.

   Another design that was made for wire saw use (given in Fig. 7.21) is a modification of the barrel holder of Fig. 4.15 and is adjusted as in Figs. 7.20.

8

Flat Grinding

GRINDING

Although grinding can be done by sliding a grinding fixture around on the surface of a flat plate to which abrasive paper has been placed, or on a piece of plate glass painted with loose abrasive in fluid, it is much faster to use a lap driven by an electric motor, as illustrated in Fig. 8.1a. The lapping plate L, generally cast iron, is caused to revolve about its axis by motor M through a belt and pulleys P and p. The lap should go no faster than the speed at which centrifugal force starts to throw off the wet abrasive. A mud pan (cross-sectional view in Fig. 8.1b) catches the drippings, thus contributing to a clean shop and also preventing abrasive from getting into the upper bearing that is between L and P.

A very useful attachment to the simple lap appears in Fig. 8.2. A post P is fastened to the frame of the lapping machine with the axis of P quite parallel to the lap axis. A channel V swings on this post. A cylinder placed in the vee should be parallel to the lap axis. If now a squared cylinder has a crystal cemented to the squared end the crystal can be ground parallel by placing the cylinder in the vee and swinging the arm back and forth over the revolving lap. The cylinder is held in the vee with the hands, which also press the cylinder against the lap. Motion of the lap should tend to drive the cylinder into the vee. A projection on top of the cylinder can stop the lowering, to produce a given thickness of crystal. The fixture of Fig. 4.15 can be used to X-ray orient a face of a crystal whereupon the crystal can be ground true to this face.

For grinding a crystal parallel and to a given thickness, the block in Fig. 8.3 is useful. One side of the crystal is flattened and cemented into the groove of the block. The exposed portion is then ground flush with the metal block. Such blocks are readily milled from single pieces of metal.

An adjustable model that does not require remilling is presented in the exploded diagram (Fig. 8.4a). Three steel blocks A, B, C, are held together by two screws S_1 and S_2. The slots in block B allow the blocks A
198

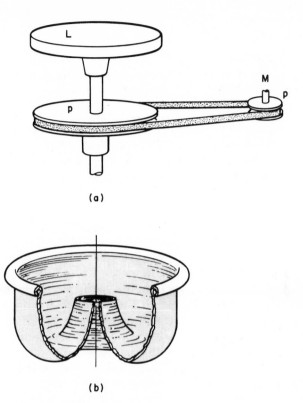

(a)

(b)

Figure 8.1 Simple, power-driven lapping machine.

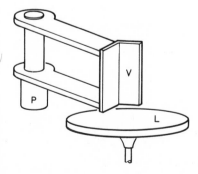

Figure 8.2 Lap attachment for squaring ends.

199

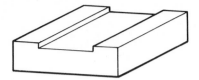

Figure 8.3 Metal block for grinding a plate to approximate thickness.

and C to project above block B by a variable amount (see Fig. 8.4*b*). Block A is tapped for screws S_1, S_2, and block C has clearance holes for these screws. The device is adjusted by means of the tool shown in Fig. 8.5. A U-shaped steel block holds an L-shaped tongue. A small block B threaded for a clamping screw is secured to the U frame by means of four rivets. The screw does not bear directly on the tongue but on a spring S, which has turned-up ends to prevent its slipping out. The spring is also slightly bent, to contact the tongue at the ends of the spring rather than at the middle. This allows the screw to be readily adjusted so that the tongue slides smoothly through the U block with a little resistance, thus remaining where it is left. When the tongue is pushed in flush to the U block, the dimension D is one inch. To set the device for producing crystal plates of thickness *t*, the tongue is pulled out a little further than the distance *t* and the screw pressure is adjusted for sliding with a slight resistance. By closing down a 1- to 2-in. micrometer, the tongue is forced in until the instrument reads $D + t$ in. The micrometer is then slacked off and removed. The screw is now tightened to lock the tongue in place. The grinding fixture of Fig. 8.4, with the two screws loosened, is inverted and placed on the adjusting fixture (Fig. 8.5). With the three elements of the grinding fixture held firmly against the reference surfaces of the adjusting fixture, the screws of the grinding fixture are tightened. Crystals are now

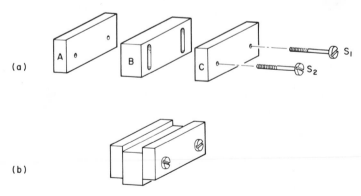

Figure 8.4 Adjustable block for approximate thickness grinding.

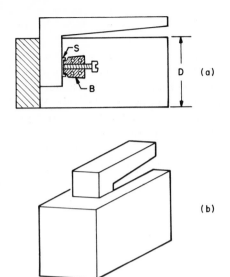

(a)

(b)

Figure 8.5   Tool for setting the block of Fig. 8.4 for desired thickness.

cemented into the "groove" of the grinding fixture and ground flush to the metal sides of the fixture. The cement should be carefully removed after each use, to prepare the device for the next setting.

An older design of grinding fixture (Fig. 8.6) consists of a heavy metal ring in which slides a solid metal cylinder. The solid cylinder has a vee-notch along the side that engages a set screw. A depth micrometer is used to set the solid cylinder at a distance t below the surface of the ring, t being the desired thickness of crystal. The set screw is tightened and the crystals cemented in. This model is a little less convenient to use than the

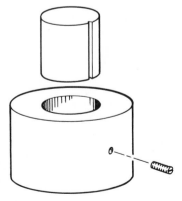

Figure 8.6   Plug and ring tool for grinding to thickness.

grinding fixture previously described. It is more difficult to set to the
desired thickness, since depth micrometers are harder to use than
standard micrometers, and the set screw tends to chew up the groove, with
the result that locking the screw changes the setting. After a little use, the
face of the hollow cylinder becomes nonperpendicular to the cylinder axis,
and the groove prevents the plug from rotating relative to the ring. When
set approximately flush, the plug and ring can be ground truly flush. If there
is now no relative rotation, the plug can be slid in the ring and can have quite
parallel planes defined by the two surfaces.

To understand the chief objection to the design of Fig. 8.6, consider
Fig. 8.7, where the plug is smaller than the hole in the ring and can tip a

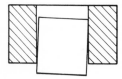

Figure 8.7   Tilt error in plug and ring fixture.

little. Tightening the set screw tends to produce this result. In a design
that overcomes this defect (Fig. 8.8), the plug rests on edges e, e so that
there is no tipping. Also a wedge-ended insert W engages the groove in the
plug, being pressed into this groove by means of the screw S. The edges e,
e are formed by boring the hole in a vertical milling machine to fit the plug
freely. A slightly smaller milling cutter is now inserted, and the worktable
is translated by a distance Δ. If the plug diameter is P and the diameter of
the second cutter is d, we have

$$\Delta = \frac{\{P - \sqrt{2d^2 - P^2}\}}{2\sqrt{2}}$$

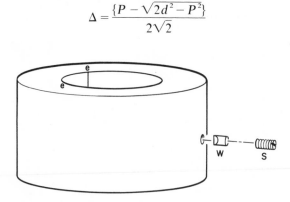

Figure 8.8   Improved plug and ring design.

It is good to take d over 7/8 of the size of P. It is wise to cut both sides as shown in Fig. 8.9. The ring is now drilled for the set screw and the second relief avoids trouble with burrs from the drilling and tapping. Finally a piece of steel is turned in a lathe to just the plug diameter, and the ring is

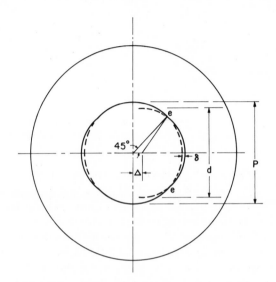

Figure 8.9 Another improved plug and ring design.

placed on this stub and locked by means of the set screw. A cut is made over the ring face and the ring is ready for use. The relief around the set screw allows the ring to slip off the stub even if the set screw has burred the stub. The depth of relief cut is

$$\delta = \Delta - \frac{P - d}{2}$$

If d is less than $P/\sqrt{2}$, then Δ has no real solution. If d is small, δ is large and the ring may be weakened.

A "GAUGE BLOCK" METHOD OF MAKING PLATES PLANE PARALLEL

With the foregoing methods we can make plates plane parallel to better than a mil, but for gauge block precision we must use gauge block methods. The simplest such method (Fig. 8.10) requires two flat ring laps. A number of crystals are carried about between these laps, the crystals being placed in holes in the central disk N. This disk or nest is moved

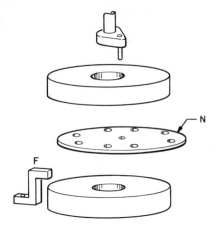

Figure 8.10 Simple "nest" method used for
paralleling plates with gauge block precision.

about by a driving pin, which is an eccentric pin on an overhead driving
shaft. Drive is often provided by a drill press that turns at about 60 rpm.
The bottom lap is constrained against motion; the top lap rests on the
crystals but is kept from moving from its position directly over the bottom
lap by three or four fingers F (only one shown). As the nest moves about,
the edges of crystals must emerge constantly from the lap edges, both the
outside edges and the edges of the holes in the laps. If the crystals do not
emerge a little, there will be a part of the lap that receives no wear.
Consequently the laps will go out of flat very quickly, as the crystal plates
will also.

With a set of similar crystals made parallel by the cruder methods
already explained, we load the nest and grind the crystals for a time.
When all the crystals are being ground—that is, when no crystal is too thin
to touch both laps at once—there is a thickest crystal of the set and a
thinnest, it being unlikely that all crystals are of the same thickness to a
high degree of precision. The deviations from the mean crystal thickness
will vary sinusoidally within the thickness spread.

If we merely continue grinding, there is no reason for the thickness
spread to decrease. However if we rearrange the crystals in the nest so
that thick and thin crystals alternate as we go around the nest, further
grinding will reduce the thickness spread, producing a new sinusoidal
distribution of deviations from the mean. There should be better ways and
poorer ways to rearrange the crystals in the nest. As the spread decreases
it becomes more difficult to identify the thickest crystal and the thinnest
crystal. We would like a transposition plan that gives the best results even
without measuring the crystals. Since the thinnest crystals and the
thickest crystals are on opposite sides of the nest, it would seem that

transposing alternate pairs of crystals at opposite ends of a diagonal would most effectively alternate thick and thin crystals. Hence we study this method.

If we consider the departure from the mean of the thickness of any crystal as a "weight," the configuration before this first transposition is out of balance relative to the center of the nest because the thicker crystals are grouped together on one side, with the thinner crystals on the opposite side of the nest. If the rate of thickness removal from a crystal is proportional to the pressure the lap exerts on the crystal, a transposition that would make the weights balance should be the desirable one. Interchanging each crystal with its opposite number accomplishes nothing. If we transpose half the crystals, leaving half unmoved, (i.e., alternate crystals are moved), we do improve the balance. For the half-set that is moved, each crystal should have an unambiguous opposite crystal that belongs to the same half-set. This can only be true if the half-set that is moved contains an even number of crystals. Hence the number of crystals in the entire set should be a multiple of 4.

To evaluate ways of transposing crystals, let us examine the model in Fig. 8.11. Here all crystals are considered to be in contact with both laps, and the laps are positioned to have the greatest separation at the top of the

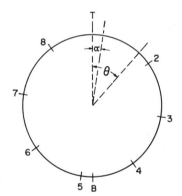

Figure 8.11 Evaluation of transposition in gauge block lapping.

diagram, the least separation at the bottom. We assume that the thickest crystal is number 1 at an angle θ to the top. If the lap separations at top and bottom differ by the amount Δ, the separation at angle θ is $(\Delta/2) \cos \theta$. Hence the thickness of the crystal in hole number 1 deviates from the mean by $(\Delta/2) \cos \alpha$. We wish to find the vertical and horizontal components of the moments about the nest center and find transpositions that minimize these moments. The moments before transposition are given in

TABLE 8.1

Number	Weight	Components[a]			
		L_v	L_h	M_v	M_h
1	$\cos \alpha$	$\cos \alpha$	$\sin \alpha$	$\cos^2 \alpha$	$\sin \alpha \cos \alpha$
2	$\cos (\alpha + 45°)$	$\cos (\alpha + 45°)$	$\sin (\alpha + 45°)$	$\cos^2 (\alpha + 45°)$	$\sin (\alpha + 45°) \cos (\alpha + 45°)$
3	$\cos (\alpha + 90°)$	$\cos (\alpha + 90°)$	$\sin (\alpha + 90°)$	$\cos^2 (\alpha + 90°)$	$\sin (\alpha + 90°) \cos (\alpha + 90°)$
4	$\cos (\alpha + 135°)$	$\cos (\alpha + 135°)$	$\sin (\alpha + 135°)$	$\cos^2 (\alpha + 135°)$	$\sin (\alpha + 135°) \cos (\alpha + 135°)$
5	$\cos (\alpha + 180°)$	$\cos (\alpha + 180°)$	$\sin (\alpha + 180°)$	$\cos^2 (\alpha + 180°)$	$\sin (\alpha + 180°) \cos (\alpha + 180°)$
6	$\cos (\alpha + 225°)$	$\cos (\alpha + 225°)$	$\sin (\alpha + 225°)$	$\cos^2 (\alpha + 225°)$	$\sin (\alpha + 225°) \cos (\alpha + 225°)$
7	$\cos (\alpha + 270°)$	$\cos (\alpha + 270°)$	$\sin (\alpha + 270°)$	$\cos^2 (\alpha + 270°)$	$\sin (\alpha + 270°) \cos (\alpha + 270°)$
8	$\cos (\alpha + 315°)$	$\cos (\alpha + 315°)$	$\sin (\alpha + 315°)$	$\cos^2 (\alpha + 315)$	$\sin (\alpha + 315°) \cos (\alpha + 315°)$

[a] L_v and L_h are the vertical and horizontal components of the lever arms of the weights, M_v and M_h are the vertical and horizontal components of the moments.

Table 8.1. To facilitate the following argument, we note a few trigonometric identities in Table 8.2.

TABLE 8.2

θ	Sin θ	cos θ
$\alpha + 90°$	$\cos \alpha$	$-\sin \alpha$
$\alpha + 135°$	$\cos (\alpha + 45°)$	$-\sin (\alpha + 45°)$
$\alpha + 180°$	$-\sin \alpha$	$-\cos \alpha$
$\alpha + 225°$	$-\sin (\alpha + 45°)$	$-\cos (\alpha + 45°)$
$\alpha + 270°$	$-\cos \alpha$	$\sin \alpha$
$\alpha + 315°$	$-\cos (\alpha + 45°)$	$\sin (\alpha + 45°)$

If we interchange crystals in holes numbers 2 and 6 and also crystals in holes numbers 4 and 8, we can make a new table for computation of moments. In this new table only the weights are interchanged, not the lever arms. However the new moments are quite different, as indicated in Table 8.3, where the vertical moments sum to zero and the horizontal moments sum to zero. The same result would follow if we had transposed all the crystals in odd numbered holes instead of the crystals in even numbered holes. If we had omitted the even numbered crystals, thus having only four crystals in the nest, and then transposing two crystals only, say numbers 1 and 5, the moments would not have summed to zero. If we had interchanged the two adjacent crystals numbers 1 and 3, we would have $M_v = 1 - \sin 2\alpha$, $M_h = \cos 2\alpha$. These vanish if $\alpha = 45°$. Determining which two crystals to interchange, however, would require measuring the crystals.

If we examine the case for 12 crystals in a nest, the diametral transposition of the half-set fails to make the sum of the moments equal to zero. In the case for 16 crystals, the moments do sum to zero.

When a great many crystals are carried in the nest, a diametral transposition of the half-set should cause the moment sum to be quite small. Other types of transposition sometimes used (e.g., across opposite edges of a square) do not seem to offer any real advantage. In the case of 12 crystals in a nest, such a transposition does not make the moments sum to zero. The analysis just presented is not quite correct. Crystals on which the lap does not bear have no "weight" until grinding establishes contact. However there is a restricted sense in which these noncontacting crystals have "weight"; that is, the amount of grinding time before contact is established is approximately proportional to the weight as previously computed.

TABLE 8.3

		Components[a]			
Number	Weight	L_v	L_h	M_v	M_h
1	$\cos \alpha$	$\cos \alpha$	$\sin \alpha$	$\cos^2 \alpha$	$\sin \alpha \cos \alpha$
2	$-\cos (\alpha + 45°)$	$\cos (\alpha + 45°)$	$\sin (\alpha + 45°)$	$-\cos^2 (\alpha + 45°)$	$-\sin (\alpha + 45°) \cos (\alpha + 45°)$
3	$-\sin \alpha$	$-\sin \alpha$	$\cos \alpha$	$\sin^2 \alpha$	$-\sin \alpha \cos \alpha$
4	$\sin (\alpha + 45°)$	$\cos (\alpha + 45°)$	$\cos (\alpha + 45°)$	$-\sin^2 (\alpha + 45°)$	$\sin (\alpha + 45°) \cos (\alpha + 45°)$
5	$-\cos \alpha$	$-\cos \alpha$	$-\sin \alpha$	$\cos^2 \alpha$	$\sin \alpha \cos \alpha$
6	$\cos (\alpha + 45°)$	$-\cos (\alpha + 45°)$	$-\sin (\alpha + 45°)$	$-\cos^2 (\alpha + 45°)$	$-\sin (\alpha + 45°) \cos (\alpha + 45°)$
7	$\sin \alpha$	$\sin \alpha$	$-\cos \alpha$	$\sin^2 \alpha$	$-\sin \alpha \cos \alpha$
8	$-\sin (\alpha + 45°)$	$\sin (\alpha + 45°)$	$-\cos (\alpha + 45°)$	$-\sin^2 (\alpha + 45°)$	$\sin (\alpha + 45°) \cos (\alpha + 45°)$

[a]Components are defined as in Table 8.1.

In summary, we have shown that to be able to transpose half a set "straight across," the number in the entire set must be a multiple of 4; but to have an automatic balance established by such a transposition, the number of pieces in the set must be a multiple of 8. Although this scheme should establish equal thicknesses with only one transposition, it is best to transpose after each periodic pause made, to check whether the proper thickness has been reached.

We have demonstrated that diametral transposition is best. If half a set is cycled around instead of diametrically transposed, the thick and thin crystals are not evenly dispersed unless the cycling involves a 180° rotation about the next center.

THE HUNT-HOFFMAN LAPPING MACHINE

A commercial machine that accomplishes a degree of transposition automatically is the Hunt–Hoffman lapping machine, represented schematically in Fig. 8.12. Here a large internal gear G engages teeth on the nests N. A spur gear g also engages teeth on these nests. Crystals placed in the holes in the nests are pressed between a lower and an upper

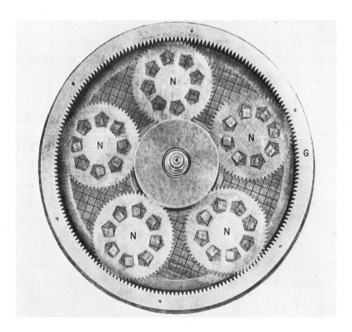

Figure 8.12 The Hunt–Hoffman lapping machine.

lapping plate. As the spur gear rotates, the nests are forced around and all crystals are ground.

FLAT LAPS

To lap crystals flat we need flat laps. A good surface grinder can make laps almost optically flat (i.e., flat to a few microns); a good lathe can do almost as well. However both machines leave a surface too rough to test by optical interference. Laps flattened by surface grinder or lathe can be improved by lapping on each other. For example, two cast iron laps can be lapped together until they "fit." The "fit" can be observed by rubbing the laps together clean and wet, later clean and dry. They soon become shiny where they touch each other, but areas that do not contact another lap remain dull. If each attains a uniform shine, the fit is good. This method should not be tried with soft laps, such as tin or lead lapping on tin or lead, because the materials gall each other. However tin or lead on cast iron or brass is acceptable. In such lapping of two laps of equal diameter, if the lapping stroke is long so that the overhang is large, say more than one-third of the lap radius, the top lap tends to become concave, the bottom one convex rather quickly. This can be used to overcome errors in flatness. An occasional position reversal tends to eliminate a progressive change.

If of three laps A, B, and C, A and B "fit," one may be concave, the other convex. Assume that A is convex. If B and C fit, then C must be convex and cannot fit A. Hence three laps that fit each other in pairs are all flat. By lapping A on B, then B on C, then C on A, then A on B, and so on, all should become flat.

A Self-Maintaining Flat Lap

In Fig. 8.13 we show a ring lap L (i.e., a lap with a central depressed area). On this lap rests a heavy steel tube T, which is restrained by the rollers R, R, R. The lap rotation causes the tube to rotate and the lap and tube to abrade each other. The lap tends to grind to a spherical shape. If the tube center is far from the lap center, the lap surface tends to become convex; if the tube center is nearer the lap center, the lap tends to become concave. Hence the rollers should be positioned to correct any departure from flatness detected in the lap surface. A proper positioning should maintain flatness once attained. The part of the lap surface not covered by the tube is available for the lapping of crystals. If an application calls for a slightly concave or a slightly convex lap, proper positioning of the tube will maintain this condition.

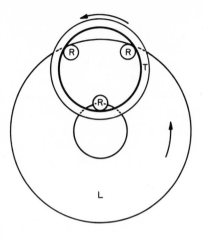

Figure 8.13 Maintaining lap flatness "automatically."

Testing for Flatness

Surfaces that are not polished enough to test by light interference can be tested by the use of a "knife edge straightedge," which is a hard steel bar (Fig. 8.14); the "edge" is cylindrical, not sharp. This allows the test to be made by holding the straightedge by hand, not necessarily strictly perpendicular to the surface. A good eye observing toward a good light source can detect a gap of about 0.5 μ on opaque laps. Glass laps pass

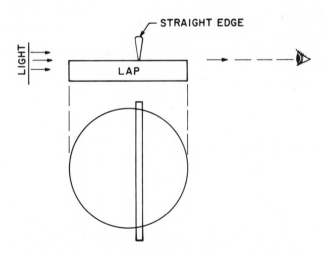

Figure 8.14 Using a knife edge straightedge to test flatness.

light scattered from the rough surface and degrade the test. A more sensitive test is to move the straightedge slightly, then look for wear marks on the lap. This is hard on the straightedge.

Light Interference Tests

Light interference tests can detect much smaller departures from flatness than can the straightedge test. Departures of 0.01 μ are detectable, and under ideal conditions this can be reduced to 0.001 μ.

In Fig. 8.15 light from a point in a distant "monochromatic" light source falls on an optical flat. The light is refracted and a wave front w_1w_2 reaches the lower surface. The light is again refracted and forms a new wavefront w_3w_4. Some of the light is reflected at w_4 and reaches an eye as shown. Part of the wave travels from point w_3 to point q, where it is reflected to point w_4 and then to the eye. Thus the two parts of the wave (i.e., the part reflected at w_4 and the part reflected at point q) arrive at the eye with a path difference $\overline{w_3q} + \overline{qw_4}$. But $\overline{qw_4} = t/\cos\alpha$, where t is the distance between the two surfaces and α is the angle of incidence of the light at the lower surface. Also $\overline{w_3q} = \overline{qw_4}\cos 2\alpha$. Hence

$$w_3q + qw_4 = \frac{t(1 + \cos 2\alpha)}{\cos \alpha}$$

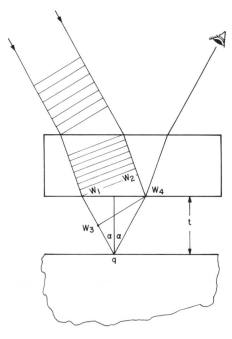

Figure 8.15 The origin of interference frin▸

and the total path difference is $2t \cos \alpha$. This is

$$\frac{2t}{\lambda} \cos \alpha \quad \text{wavelengths}$$

Because of a half-wave phase change caused by reflection at point w_4, we see that there will be destructive interference between the two parts of the wave if $2t \cos \alpha$ is an integral number of wavelengths, constructive if it is some integer plus a half wavelength. Hence if the air gap tapers a little, we will see light and dark fringes crossing the surface. If α is kept small, say less than 15°, the error in neglecting it is only a few percent and we can assume that one fringe corresponds to a $\lambda/2$ difference in air gap.

With an optical flat placed on a reflecting lap, we see concentric rings if the lap has a spherical surface. If the radius of curvature is very great there will be a few widely spaced fringes, but this does not indicate whether the lap is concave or convex. However if we press the optical flat down on the near side, as in Fig. 8.16, the fringes can be thought of as showing a profile of the lap: a convex lap, a concave lap, and a combination often seen on laps are illustrated. By placing a straight line across one of the fringe patterns, we can tell the departure from flatness. In this way we see that the lap of Fig. 8.16c is in error by less than a fringe (i.e., less than half a wavelength). For light of wavelength 5000 Å this is an error of less than 2500 Å (i.e., less than 10 μin.).

There are many sources of monochromatic light. Most give several wavelengths, but one color can be separated by light filters. Mercury, sodium, and thallium discharge tubes are available. Some fluorescent lamps give light sufficiently monochromatic for these flatness tests. Even what appears to be a single color may have a broad wavelength band or several sharp lines close together. Sodium with two lines close together ($\lambda_1 = 5889.95$ Å, $\lambda_2 = 5895.92$) has one more λ_1 wave than it has λ_2 waves in a path length of 0.05817 cm. Hence in a distance of 0.145 mm the two waves are out of step by a half a wave, and the dark fringes of one are superimposed on the bright fringes of the other. The result is the apparent disappearance of all fringes for this path difference, since the two components are of nearly the same intensity.

TIPPING DUE TO OFF-CENTER PRESSURE IN DISK GRINDING

In grinding disks on a lap it is often necessary to apply the pressure off center to hasten the grinding at one edge, perhaps because although the disk is thickest at this edge, the requirement is for a parallel disk. If the pressure is applied too far from the center, experience shows that the opposite edge may actually lift from the lap, producing a nonflat surface

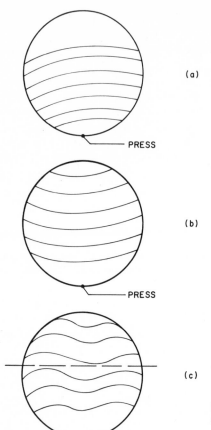

(a)

PRESS

(b)

PRESS

(c)

PRESS

Figure 8.16 Determining the nature of departures from flatness. (*a*) Convex lap, (*b*) concave lap, (*c*) combination frequently observed on laps.

on the disk. The problem we seek to answer here is "How far from the center may the force be applied?" We assume the plate is thin enough that the overturning torque is small and that the plate is thick enough not to bend.

Case I. A Circular Disk

Consider a thin disk being ground on a large lap (Fig. 8.17). The disk is forced against the lap by means of a pin in a small pit, off center by amount h. The disk is moved over the surface of the rotating lap so as to cover all parts of the lap equally. This helps to keep the lap flat. Also the grinding is assumed to proceed in all directions, ensuring that there is no perpetually leading edge to be ground more than a trailing edge. We

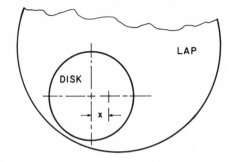

Figure 8.17 Analysis of local pressures in lapping a round disk.

assume that under these conditions the rate of removal of material by the lap from any point on the disk is proportional to pressure between disk and lap at that point. If a rigid plate be pressed against an elastic substrate the depression is linear. Hence by Hook's law the pressure is a linear function of the distance x (Fig. 8.18), that is,

$$p = ax + b \tag{8.1}$$

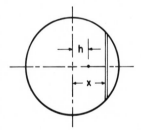

Figure 8.18 Further analysis of local pressures in a round disk.

Since the mean pressure is b, it follows that

$$F = \pi R^2 b \tag{8.2}$$

Taking moments about the origin, we have

$$hF = 2 \int_{-R}^{R} x(ax + b)\sqrt{R^2 - x^2}\, dx$$

whence

$$a = \frac{4hF}{\pi R^4} \quad \text{and} \quad p = \frac{F}{\pi R^2}\left(\frac{4hx}{R^2} + 1\right)$$

This vanishes at $x = -R$ if $h = R/4$ and is negative if $h > R/4$. Hence if $h > R/4$, an edge of the disk tends to lift from the lap and a flat surface is not produced on the disk.

Case II. A Rectangular Plate

We assume

$$p = ax + by + c \qquad (8.3)$$

Again the average pressure is c, giving

$$c = \frac{F}{WL}$$

and if the pressure point Fig. 8.19 is off center by amounts h in the x direction and k in the y direction, the amount about the y axis is

$$Fh = \int_{x=-L/2}^{L/2} \int_{y=-w/2}^{w/2} x(ax + by + c)\, dy\, dx$$

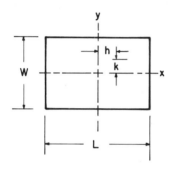

Figure 8.19 Analysis of local pressures in a rectangular plate.

This allows us to write

$$Fh = \frac{awL^3}{12}$$

Similarly, moments about the x axis give us

$$Fk = \frac{bW^3L}{12}$$

Hence we write

$$P = \frac{12Fhx}{WL^3} + \frac{12Fky}{W^3L} + \frac{F}{WL}$$

For P to vanish at the corner $x = -L/2$, $y = -W/2$, we have

$$\frac{6h}{L} + \frac{6k}{W} = 1 \qquad (8.4)$$

Thus if this corner is not to lift, the force must be applied below the diagonal line (Fig. 8.20). If *no* corner is to lift, the pressure point must be inside the diamond (Fig. 8.21).

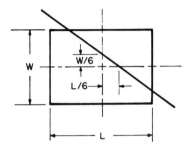

Figure 8.20 The boundary between pressure points that cause no negative pressures.

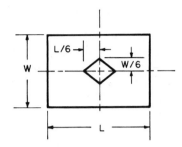

Figure 8.21 Pressure inside the diamond causes no negative pressure.

Case III. Interrupted Plate

A case of great practical interest is that in which there are gaps in the plate, illustrated in Fig. 8.22. Four feet are arranged around a central block. For definiteness we will assume the blocks have sides one-fifth the

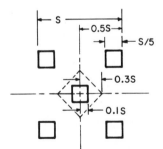

Figure 8.22 Limiting lines for no negative pressures in the lapping of five square blocks.

side of the circumscribed square. If the pressure point is displaced along one axis as in Fig. 8.22, we have

$$p = ax + c \qquad \text{where} \quad c = \frac{5F}{S^2}$$

Taking moments yields

$$Fh = \frac{2S}{5} \int_{x=-0.5S}^{-0.3S} x(ax+c)\,dx + \frac{S}{5}\int_{x=-0.1S}^{0.1S} x(ax+c)\,dx$$

$$+\frac{2S}{5}\int_{x=0.3S}^{0.5S} x(ax+c)\,dx = \frac{0.394S^4 a}{15}$$

$$a = \frac{15Fh}{0.394S^4}, \qquad p = \frac{15Fh}{0.394S^4}x + \frac{5F}{S^2}$$

which vanishes for $x = -0.5S$ if $h = 0.262S$. Hence the pressure point can be moved out farther in the case of the interrupted plate. The dotted line of Fig. 8.22 is the limiting line.

PRODUCTION OF SONAR CRYSTAL PLATES

In the production of ammonium dihydrogen phosphate plates for World War II sonar systems, the paralleling and edging were done on dry sanding belts. The crystals directly from the growing tanks were sliced up with abrasive saws, using propylene glycol and water saturated with the salt. The method of orientation was described in Chapter 5. The final plates had their thickness directions along the crystal prism axis. The plate length and width directions were 45° to the prism faces. An inexpensive drafting machine was used to pencil out plate profiles (Fig. 8.23). Without this laying out it would be easy to grind the first corner so far that a full-sized plate could not be completed. The marked crystals were then belt ground on two adjacent corners (Fig. 8.24), the shallow cut being made first. Each cut took a few seconds only. With these two new surfaces as reference faces, the two paralleling cuts were made on a similar belt grinder in which slide D was replaced by a new slide (Fig. 8.25). The plate with two finished edges was first placed on the left side of the slide and the slide pushed forward until the stop screw S struck the base B (Fig. 8.24). This step finished the width to specifications. The plate was then placed in the right notch and pushed against the grinding belt until the stop S again set the dimension L this time the length. For finishing the thickness dimension to specifications, a third slide was used (Fig. 8.26, seen from the belt side), but only for laboratory-scale production.

For production-scale finishing of the thickness, an entirely separate machine was built (Fig. 8.27). A cast iron plate P about a foot in diameter rode on two phosphor bronze blocks B, B. Each block had a groove communicating with exhaust tubes that were connected to a vacuum pump. The disk P had a number of holes that rode over the vacuum

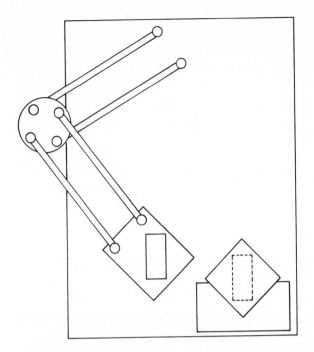

Figure 8.23 Laying out ADP crystals.

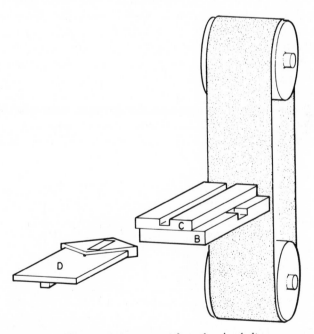

Figure 8.24 Grinding ADP crystals by means of an abrasive belt.

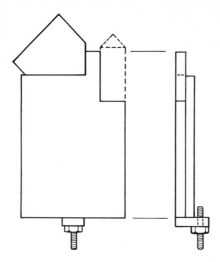

Figure 8.25   Continuing the ADP cutting to orientation.

groove in blocks B. Over each of the bronze blocks was a small belt grinder: the lower pulley of each was a true cylinder, the upper pulley was crowned. Only one belt grinder appears in Fig. 8.27. Crystals placed over the holes of plate P slowly passed under one belt, were turned over and then passed under the second belt. As a crystal approached a belt, the hole over which it was placed entered the region of the vacuum groove; thus the crystal was sucked down, to be held firmly against the forces of the grinder. Obviously the second grinder was set for the final thickness specification; the first belt was set a little higher.

Belt grinders such as that in Fig. 8.24, had hardened steel backing plates that were water cooled where fast use dictated this. The belts were ordered with the specification that the belt joint be thinner than the rest of the belt. If the belts did not flex as they passed around the pulleys, they

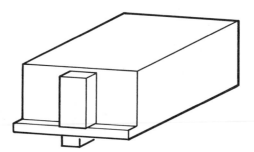

Figure 8.26   Grinding ADP to final thickness.

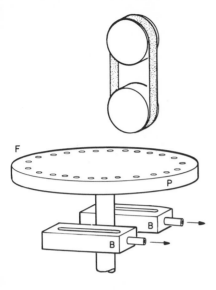

Figure 8.27 Automatic grinding of ADP crystals to required thickness.

would have quickly become clogged with crystal dust. However the flexing, aided by centrifugal force, kept the belts quite clean. A vacuum dust removal system was required to keep the ambient atmosphere respectable. In later experiments on ammonium dihydrogen arsenate the belts were run wet to keep the arsenate dust out of the atmosphere. The belt speed was kept low enough to prevent a mist of arsenate-bearing fluid from poisoning the atmosphere.

The Diamond Fly Cutter

For making thin slabs of Rochelle salt, a high-speed diamond-tipped fly cutter (Fig. 8.28) was used. Although this device performed very well with Rochelle salt, it chipped the edges of ammonium dihydrogen phosphate. A diamond D carried on an arm A by the shaft S is turned by an air blast against a notched wheel W. The notches, merely indicated on the wheel, appear in more detail in the inset, right center. The exhaust air goes up a stack P, into which a silencer is built. The plate RS was stuck down on the bed by a little grease. The bed was carried by a lathe cross slide (not shown). The diamond shaft turned faster than 35,000 rpm, and because of this speed the crystal received a very smooth finish even when the feed handle was turned as rapidly as it could be turned by hand.

The diamond arm had to be very carefully balanced. The arm was screwed onto a balancing shaft as shown in the inset, lower left. The shaft was placed on a pair of parallel knife edges, which were placed in turn on a level surface plate. The arm was filed until the arm and balancing shaft

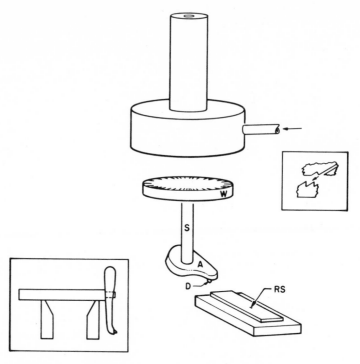

Figure 8.28 High-speed milling of Rochelle salt surfaces.

would remain in any position on the knife edges. Figure 8.29 is a photograph of the machine.

Long Slender Rods

If the end faces of long slender rods are required to be flat and parallel to a few seconds of arc (as is the case for solid state laser rods), not all the previous methods are applicable. To meet a requirement for a great many such rods all alike, "bundles" could be prepared by cementing rods into holes in cylindrical metal blocks, as in Fig. 8.30. Eight such blocks could then be ground in a nest as in the gauge block method. High precision is thus assured if the laps are kept in good shape.

However experimental work sometimes calls for only one rod, and material for more rods may not be available. To satisfy this requirement we cement the rod into the center of a steel block so that both ends protrude. Each protruding end is surrounded by four feet that grind away as the rod grinds away. The feet stabilize the assembly against tipping due to the forces of grinding. In Fig. 8.31 the block is supported by a gimbal

Figure 8.29 Pneumatically driven, diamond tool crystal surfacer.

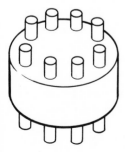

Figure 8.30 A batch of laser rods, polishing the ends parallel.

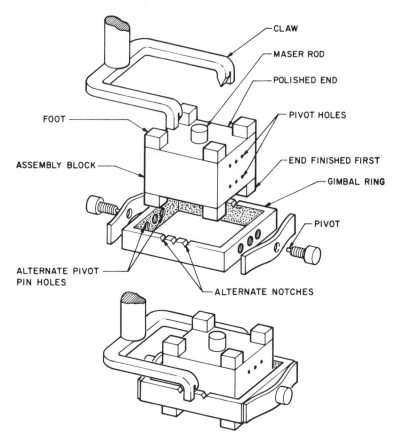

Figure 8.31   Fixture for paralleling one laser rod.

ring, which is guided by a claw that engages notches in the gimbal ring. There are three pairs of notches in the ring. The central pair are for use if grinding is supposed to proceed at the same rate on all four feet. However if one side needs to wear away faster than the other side, a pair of the alternative notches is selected. There are three pairs of pivot holes also at each end of the block. These also allow for grinding one edge faster than the other.

The steel block is optically polished on one end. Grinding and polishing the rod end opposite the polished face of the block permits comparison of reflections from the block with reflections from the rod end. This is done in an autocollimator (Fig. 8.32). If the image of the target formed by light

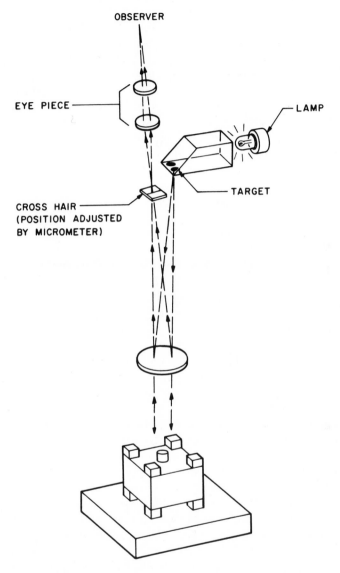

OBSERVER

EYE PIECE

LAMP

CROSS HAIR
(POSITION ADJUSTED
BY MICROMETER)

TARGET

Figure 8.32 Autocollimator for testing the parallelness of laser rods.

reflected from the rod end coincides with the target image formed by light reflected from the block, the freshly polished face is parallel to the block face within the resolution of the autocollimator. The block is best prepared by drilling and reaming the central hole first, then lightly pressing in a true arbor several inches long. If a lathe chuck holding the block is adjusted until two spots several inches apart on this arbor run true by dial indicator test, the hole in the block runs true to the lathe axis within a fraction of a minute of arc. A light cut taken on the end renders this end perpendicular to the hole. We now place this face down on a good surface grinder and grind the other end of the block parallel to the end that was trued in the lathe. Using the gimbal rig (Fig. 8.31), we now lap this ground end on a flat tin lap charged with fine emery or diamond. If we grind the minimum necessary to give good reflections, we will not lose much of our fractional minute accuracy.

This grinding is best done on an optical machine like the one in Fig. 8.33. Here a sweep arm holds the claw that engages the gimbal ring notches. The arm can be given any length of stroke likely to be needed and can center the stroke wherever desired. The claw can be removed and a simple ballpoint pin substituted for the grinding or polishing of disks such as laps or samples cemented to plates. The ball enters a shallow hole at the disk center, driving the disk but leaving it free to rotate and tip enough to follow a wobbling lap.

Returning to the rod polishing, we test the flatness of the polished end by means of an optical flat and monochromatic light. This is best done in a Fizeau interferometer (Fig. 8.34). If the interference fringes are parallel and equally spaced, the rod end is flat. If the fringes are curved and an imaginary chord across one fringe is tangent to the adjacent fringe, the "elevation" of the point of tangency departs from the elevation of the chord ends by half a wavelength. If the rod is much out of flat, the autocollimator image from the rod will be fuzzy.

Having satisfactorily finished one end of the rod, we turn our attention to the other end. We measure the distance between flat surfaces at opposite ends of the block (i.e., the length parallel to the rod length). If these four lengths are unequal, we grind, using alternate pivot holes and notches until the lengths are equal. If diagonally opposite lengths are 1.5 in. apart and differ by, say, 0.2 mil, the rod ends are out of parallel by about 22 seconds of arc. Polishing can now commence, and the autocollimator can be used to determine parallelness. This is done by comparing the image position for reflections from the rod with image position for reflection from the base on which the block assembly is placed—the autocollimator base.

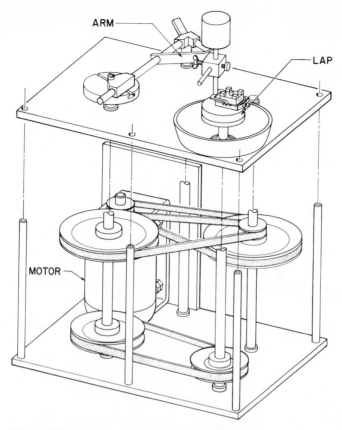

Figure 8.33   Lapping machine made to lap laser rods.

In this work the laps must be maintained very flat. A high-quality lathe can produce a good enough flat. It may be necessary to shim the carriage to ensure that the cut is not conical.

A composite vee block (Fig. 8.35) can be used to grind a range of larger cylinder ends parallel and perpendicular to the cylinder axis. The composite is made up of a frame F and two vee segments V_1 and V_2. The composite is clamped firmly together by means of two socket-head screws S, S. Set screws (not visible here) come up from below and press against a pressure pad that in turn presses against the cylinder, thus locking it in place. Cement can be used to further lock the crystal in place. Since the composite can be disassembled the vee segments can be scrubbed clean and lapped flat.

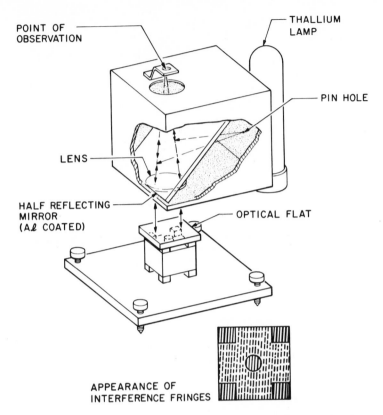

POINT OF OBSERVATION

THALLIUM LAMP

PIN HOLE

LENS

HALF REFLECTING MIRROR (Aℓ COATED)

OPTICAL FLAT

APPEARANCE OF INTERFERENCE FRINGES

Figure 8.34 Fizeau interferometer for testing flatness of laser rod ends.

MAKING CRYSTAL PRISMS

Measuring the refractive indices of crystals requires prisms, and only for cubic crystals is the orientation of a prism unimportant. For uniaxial crystals the c axis should lie in the prism bisecting plane (Fig. 8.36). It is an added convenience if c is either parallel to the prism edge or perpendicular to it.

For orthorhombic crystals we need two prisms. For one prism, two of the axes a, b, c must lie in the bisecting plane; for the other prism two of the axes a, b, c must lie in the bisecting plane, but not the same two axes. Again it is convenient if one axis is parallel to a prism edge, another perpendicular to it. For monoclinic crystals a and b must lie in the bisecting plane of one prism, b and c must lie in the bisecting plane for the other prism.

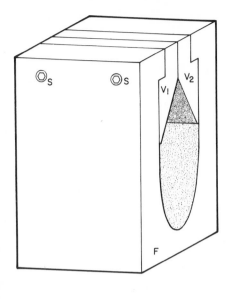

Figure 8.35 Device for squaring crystal cylinders.

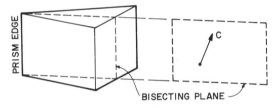

Figure 8.36 Orientation of prisms for determination of refractive indices.

Such prisms are produced by means of X-ray orientation. After fine grinding the transmitting faces, a few strokes on a tin lap charged with fine emery renders the transmitting faces sufficiently reflecting to permit optical polishing without losing the orientation. This is done with the implement of Fig. 8.37. Here a gimbal ring R surrounds a metal plate P. On the bottom of the plate there are several supporting plates of material having about the same hardness as the prism material. The secondary plates have been ground and polished after cementing onto the metal plate so that their exposed faces are coplanar. The X-ray corrected prism is cemented onto the mounting plate MP and the dovetail S is inserted into its slot. Examination of the reflection of a straight line with grazing incidence reveals that the portions of the line image are not continuous but offset portions, unless the prism surface is parallel to the plane of the supporting blocks (see Fig. 8.38). It is necessary to sight twice, the second

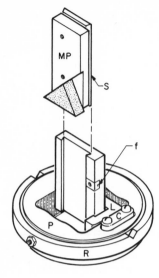

Figure 8.37 Device for grinding and polishing two faces of crystal prisms.

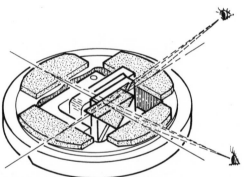

Figure 8.38 Sighting in to adjust the prism processing device of Fig. 8.37.

sighting being at right angles to the first one. The piece L is a bracket that supports a short shaft that can be locked by tightening the screws of the clamp c. Another such clamp locks another short shaft, but this is hidden by the body of the dovetail receiver. By means of these two shafts the lower prism surface can be made parallel to the supporting block surfaces and locked in place. The dovetail slide can now be adjusted to make the prism surface flush with the surfaces of the supporting blocks. The assembly now goes onto the lapping machine (Fig. 8.33), the claw engaging the notches in the ring R (Fig. 8.37). After fine grinding and polishing, the prism is remounted to process the other transmitting surface of the prism. An autocollimator can be used instead of the naked eye sighting described earlier.

9

Making Cylindrical Surfaces

THE MAKING OF CIRCULAR CYLINDERS

There are several common methods for making circular disks (i.e., very short cylinders). The simplest for the quick, offhand manufacture of disks is to cement a plate to a cylindrical metal piece, then grind the plate down to the metal on a power-driven lap (Fig. 9.1). The work can be considerably lessened by sawing off corners before grinding. A higher degree of

Figure 9.1 Grinding a circular disk.

roundness is obtained by finishing the disk to size, employing a lapping fixture (Fig. 9.2). The approximately round disk is almost surrounded by a metal strap, which can be pinched together with the fingers. The disk, cemented to the end of a metal cylinder, is rotated and abrasive is applied to the inside of the strap. As the disk and strap wear away, the fingers, pressing at points P, P close the strap. With a little care the disk can be made round to a fraction of a mil.

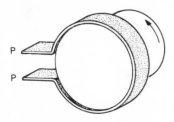

P

P

Figure 9.2 A more precise method of grinding a circular disk.

231

The same method works well with long cylinders. A long cylinder can be started by making a square prism, then placing the prism in a metal block into which a 90° groove has been milled, as in Fig. 9.3. If S is the side of the square prism then the depth of the groove should be $D = S(\frac{1}{2} + 1/\sqrt{2}) = 1.2071S$. Its width should be twice this: $W = S(1 + \sqrt{2}) = 2.4142S$. The crystal is cemented into the groove and ground flush to the metal. This is repeated around to give quite an accurate octagonal prism. If a surface grinder is available it should be just touched to the prism edge, then the wheel is lowered in steps by amount $d = S(1/\sqrt{2} - \frac{1}{2}) = 0.2071S$ when the flat should have width $w = S(\sqrt{2} - 1) = 0.4145$. It is accurate enough for most purposes to grind off one-fifth S from each edge, at 45°, leaving a width $w = 0.4S$. In making long cylinders from such long octagonal prisms, a wider strap than that shown in Fig. 9.2 should be chosen. Less breakage occurs with fragile material if the prism is rotated by means of a flexible drive. Such a drive is given by a flexible tube—rubber or tygon, for example. The tube should be sized to require force to press the prism a quarter of its length into the tube. The other end of the tube is forced onto a metal rod driven by a lathe or drill press. During the grinding the prism is frequently reversed end for

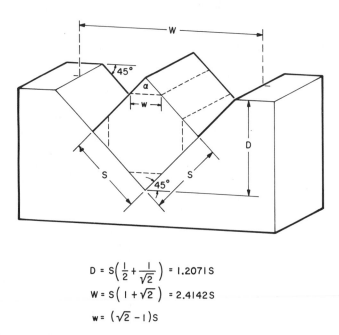

$$D = S\left(\frac{1}{2} + \frac{1}{\sqrt{2}}\right) = 1.2071S$$

$$W = S\left(1 + \sqrt{2}\right) = 2.4142S$$

$$w = \left(\sqrt{2} - 1\right)S$$

Figure 9.3 Making an octagonal prism preparatory to grinding a circular cylinder.

end, to make use of its entire length. The strap is slid back and forth along the exposed length of the prism as the prism rotates. A screw through the strap ends can apply the closing pressure. This helps make the diameter the same all along the finished length.

Returning to short circular cylinders, that is, disks, larger scale operations can produce disks by trephining, often called ("trepanning"). The only equipment required is a slow-speed drill press. The method is represented schematically in Fig. 9.4. The trephining tool has its cutting edge nicked as shown to give the abrasive entry points. The rotating tool is pressed against the surface after the surface has been flooded with a slurry of abrasive. An oil-bound clay dam around the area keeps the slurry where it is needed. After the cutting sound changes from a coarse crunching sound to a hissing sound, the abrasive is spent. To replenish the tool, it is lifted a little and is again brought down on the work. The most common mistake is the too infrequent lifting of the tool for abrasive

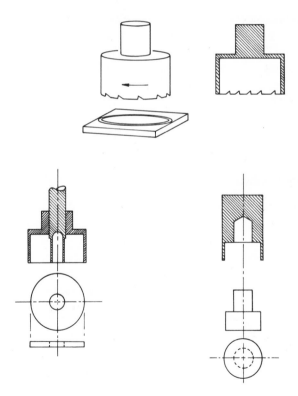

Figure 9.4 Trephining disks.

replenishing. Very deep notches allow for a longer dwell time. The actual times depend on the hardness of the crystal and the size of the crystal. Of course the crystal has previously been cemented to a scratch layer and a holding block. Without the scratch layer the underside of the crystal may chip badly as the tool bursts through. If the crystal is mounted on a metal block without the scratch layer, and if the metal cuts slowly, a flash may be found around the disk when the crystal is dismounted. The scratch layer should cut as readily as the crystal.

As the drawings in the lower portion of Fig. 9.4 indicate, more complicated structures with circular symmetries are readily produced. (The work in the lower portion was produced by the tool appearing above it.) Notice the two-piece construction of the structure on the left. This is done to make its fabrication easier. However this design has the advantage that if the two cutting edges wear at different rates they can be readjusted to the same level. The cutting edges and their backing should be kept as thin as is consistent with sufficient rigidity. Thick walls require more abrasive, more cutting time, more power. Tool wear tends to cause the work to taper a bit, the corners intended to be sharp become rounded, and a second finishing tool should be employed after the roughing out is done. Mild steel is the most practical substance from which to make the tools, but spring-tempered tool steel would last longer. If the tools are made of copper and charged with diamond, however, they should be used at a much higher speed of rotation.

Somewhat resembling trephining is "cookie cutting." In this process a die is repeatedly lowered forcibly against the workpiece, then backed off. This is done at frequencies in the thousands per second. An abrasive slurry floods the workpiece. At each backing off fresh abrasive is sucked under the die. At each descent some abrasive is trapped under the die and is crunched into the crystal. It pulverizes a certain amount of crystal at each descent, and much of the debris is washed out at the next descent. Since the cookie die does not need to have circular symmetry, quite complex shapes are made. Medallions are carved in stone commercially with such tools. Figure 9.5 shows some of the details.

For continuous production of circular disks, an "edger" can well be employed. This machine is like that used in making the fancy shapes of spectacle lenses. The workpiece is clamped between two soft-pad-lined shaft ends (left, Fig. 9.6). If the two shafts are not accurately coaxial or are driven at different speeds, the workpiece will shift around in its soft pads SP and give poor results. If one shaft only is driven, and the other turns readily on ball bearings and coaxiality is accurate, good results follow. The arm A is slowly fed in about the pivot P as the cutting proceeds. The work must be traversed across the grinding wheel W to

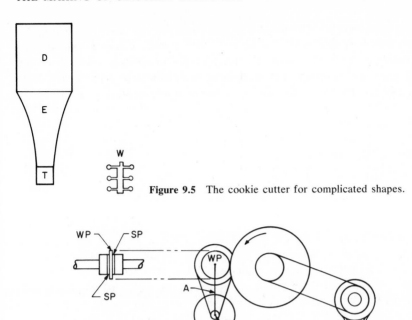

Figure 9.5   The cookie cutter for complicated shapes.

Figure 9.6   An edge grinder.

distribute the wear. A coolant fluid is constantly played on the grinding wheel near its point of action.

In the making of many disks it is better to cement many of the roughed out disks together to make a long cylinder. Figure 9.7 illustrates a tool to facilitate getting the disks normal to the axis of the cylinder and the disk centers at the cylinder center. Two short cylinders C, C with centered conical pits at one end, the other end very square to the axis, are used to press the workpieces together in a vee block. A clamp is then tightened to

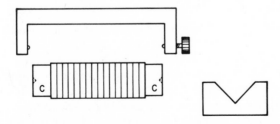

Figure 9.7   Cementing a group of disks preparatory to edging or centerless grinding.

secure the assembly which can be lifted from the vee block and heated for cementing. It is permissible to have the cylinders C, C a little smaller than the finished "cylinder" will be, shims in the vee block assuring proper centering. The cylinder is then ground in a centerless grinder (Fig. 9.8), which is commercially available. In Swiss-made centerless grinders for

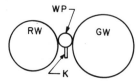

Figure 9.8   The centerless grinder.

the manufacture of watch jewels, the grinding is done by diamond-charged copper cylinders. In Fig. 9.8 the workpiece is between the grinding wheel GW and the regulating wheel RW but above the line of centers of GW and RW. The knife K keeps the work from being forced between the two wheels and assures that the center of WP is above the line of centers of GW and RW. To reduce the size of WP, the wheel RW can be moved in toward GW.

We have made and used a very small centerless grinder in which the wheels were replaced by smooth steel cylinders. The grinding was done by an oil slurry of abrasive, producing cylinders of fractional millimeter size from soft, water-soluble crystals. A simple device for making cylinders is the nest shown in Fig. 9.9. Octagonal prisms are placed in the

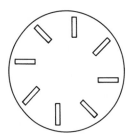

Figure 9.9   Grinding a batch of small circular cylinders.

slots, and the nest is made to rotate about its center by rotating the upper of two ring laps between which the nest is placed. Since the slots are not strictly radial, the prisms must slide as well as roll, and this sliding grinds the prisms "round." Frequent transposition of half the pieces assists in making the pieces all alike. Sometimes having one plate made of rubber forces the rotation and grinding by the other plate, which can be an abrasive surface. We have even used abrasive paper cemented to one lap, the other being rubber. However these cylinders are subject to the cylindrical paradox.

The importance of the nonalignment of the center of GW, WP, and RW can be appreciated by a study of a cylindrical paradox.

THE CYLINDRICAL PARADOX

Consider three noncollinear points as the apices of a triangle (Fig. 9.10). We call the one at the most acute apex 1; the others are 2 and 3. We draw the lines 12, 23, 31, but distinguish the two ends of a line by a reversal of numerals (i.e., 12 is the other end of 21). With a center at point 1 and an arbitrary radius r_1, we draw an arc between lines 21 and 31. Then with a

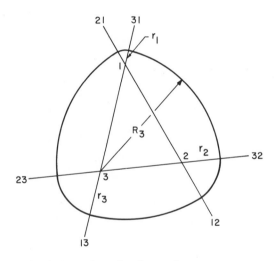

Figure 9.10   The cylindrical paradox using three points.

center at point 3 and radius R_3, we draw an arc from line 31 to line 32. This arc is a continuation of the previous arc if $R_3 = r_1 + d_{31}$, where d_{31} is the distance between points 1 and 3. We draw it as such. With a center at point 2 and a radius $r_2 = R_3 - d_{23}$, we draw an arc from line 32 to line 12. This arc is a continuation of the previous arc. With a center at point 1 and a radius $R_1 = r_2 + d_{21}$, we draw an arc from line 12 to line 13. Once again, the new arc continues the previous arc. With a center at point 3 and a radius $r_3 = R_1 - d_{31}$, we draw from line 13 to line 23 an arc that continues the previous arc. With a center at point 2 and a radius $R_2 = r_3 + d_{23}$, we draw an arc from line 23 to line 21. This arc continues the previous arc and joins the first arc smoothly because it can easily be demonstrated that $R_2 = r_3 + d_{23} = r_1 + d_{12}$. This is made clearer in Table 9.1.

TABLE 9.1   TABLE OF RADII

r_1 given	
$R_3 = r_1 + d_{31}$	
$r_2 = R_3 - d_{23}$	$= r_1 + d_{31} - d_{23}$
$R_1 = r_2 + d_{12}$	$= r_1 + d_{31} - d_{23} + d_{12}$
$r_3 = R_1 - d_{31}$	$= r_1 - d_{23} + d_{12}$
$R_2 = r_3 + d_{23}$	$= r_1 + d_{12}$

From the foregoing tabulation it follows that

$$r_1 + R_1 = r_2 + R_2 = r_3 + R_3 = 2r_1 + d_{12} - d_{23} + d_{31}$$

Let us now use parallel-jawed calipers to measure the "diameter" of our composite curve with one jaw in the arc between 21 and 31, the other on the arc between 12 and 13. We find the answer to be $r_1 + R_1$, whereas if we measure with one jaw on the arc between 32 and 12, the other on the arc between 23 and 21, we find $r_2 + R_2$. But we have proved that these are equal. Hence all "diameters" are the same, but the figure is obviously not a circle. We specified "parallel-jawed calipers" to measure these diameters for a reason: if one jaw is a vee block, as in Fig. 9.11, the "diameters" do not measure the same all around. This paradox is not limited to three-point composite curves but extends to any odd number of points. A construction for five points appears in Fig. 9.12.

Returning to the centerless grinder, if the centers of GW, WP, and RW

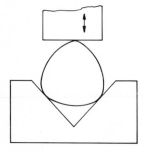

Figure 9.11 Disk that measures as a perfect circle if measured between parallel surfaces but not if measured as shown.

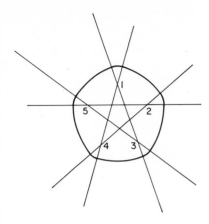

Figure 9.12   The cylindrical paradox with five "centers."

are on a straight line, a cylinder with a cross section like that of Figs. 9.10 or 9.12 would satisfy the mechanical conditions; that is, it could contact the two wheels and the knife for any position of the work. Hence true roundness would not be produced. By grinding with the centers out of line, we can achieve true roundness.

PARTIAL CYLINDERS

One way to obtain a partial cylinder such as a cylindrical lens is to cut up a cylinder. Another way is to cement strips together, grind a cylinder from this composite, and finally uncement the assembly. Although these methods may be satisfactory for cylinders of small radius, they will not do for partial cylinders of large radius, which require that we grind on curved laps.

The machine in Fig. 9.13 is made for working cylindrical surfaces. A geared motor at M carries a two-step, belt-driven pulley P. This pulley drives two adjustable-crank rotors R1 and R2, which cause the arms A1 and A2 to oscillate about the pins P1 and P2 with adjustable amplitudes. In Fig. 9.14 a saddle piece SP sits on top of arm A1, slotted to allow it to be slid along A1 to a desired position and locked with a screw into a hole (h in Fig. 9.13). The right end of SP has a hole for a large thumb screw that engages a hole in the mounting plate MP. The lens GL is cemented to the mounting plate. The lap LP, which rests on the lens, has three shallow holes on its upper surface. The central hole commonly engages a driving pin DP, but if one end of the lens is too thick, one of the other holes allows greater pressure to be applied to the thick end.

In Fig. 9.15 the driving pin DP is held in the arm extension AE by means

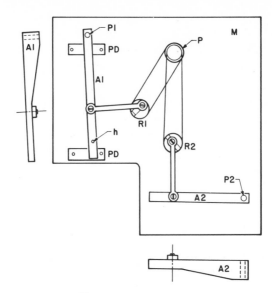

Figure 9.13   The two-arm lapping machine.

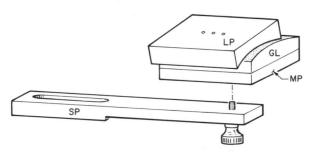

Figure 9.14   Making a cylindrical lens using the two-arm lapping machine.

of a set screw. The central vertical rod is used for weights to hasten grinding. At the right end of AE are pivot holes PH, which engage pivot screws PV in the bracket BK. Bracket BK fits on top of arm A2 (Fig. 9.13) and is slotted to ensure the accurate positioning of motion of lap over lens. When the screw in R1 is loosened, the dovetail piece can be slid out of its slot and arm A1 can be slid off pin P1. A rotating shaft (Fig. 9.16) slips into place; the bearing blocks of Fig. 9.16 are held in place by screws into the pads PD of Fig. 9.13. The belt to R1 (Fig. 9.13) is replaced by a belt to the pulley on the shaft (Fig. 9.16). In Fig. 9.17 six cylindrical lenses are being ground in this way.

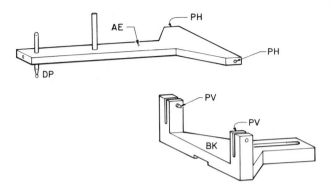

Figure 9.15 Detail of driving arm of two-arm lapping machine.

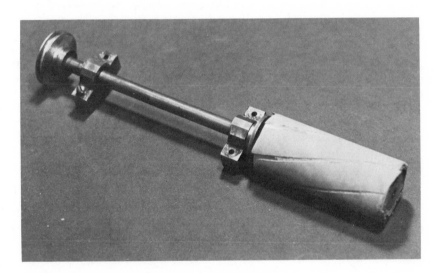

Figure 9.16 Polishing lap for a concave cylindrical surface.

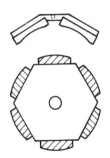

Figure 9.17 Making six cylindrical lenses.

241

Laps of very small or very large radius present problems. For small radii an end mill can be used in a milling machine to produce a groove that has the lap radius. Milling cutters of special radius can be made by turning a cylinder on a tool steel rod, then rounding the end into a hemisphere with a file guided by a template. Half the end is cut away, as in Fig. 9.18.

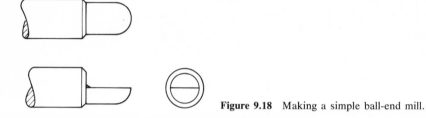

Figure 9.18   Making a simple ball-end mill.

The tool is then hardened and tempered to a straw color. A better (i.e., freer) tool can be obtained by making the cylinder 1% oversize and cutting away more than half, leaving 43% of the original diameter. The end is then bent to center, as in Fig. 9.19.

Figure 9.19   Better design of the simple ball-end mill.

Commercial cutters come in various sizes and are designated by the diameter, not by radius of cut. These can be used to approximate a cylindrical groove of any desired radius if the next smaller size than the required groove "diameter" is chosen. Referring to Fig. 9.20, a tool that

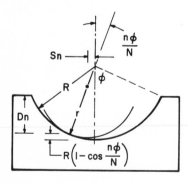

Figure 9.20   Making a greater radius with a smaller radius tool.

cuts a radius r is used to produce a groove of radius R and included angle ϕ.

After making the central cut to a depth D_0, where

$$D_0 = R(1 - \cos \phi)$$

we make $2N$ other cuts at different depths and offsets. The depth of the Nth cut is

$$D_n = D_0 - \Delta \cos \frac{\phi N}{N}$$

where $\Delta = R - n$. The offset for the two nth cuts is

$$S_n = \pm \Delta \sin \frac{n\phi}{N}$$

This leaves N ridges of thickness ε on the surface:

$$\varepsilon = R - \Delta \cos \frac{\phi}{2N} - \sqrt{R^2 - 2RD + \left(\Delta \cos \frac{\phi}{2N}\right)^2}$$

which is approximately equal to

$$\varepsilon \sim \Delta \left(1 - \cos \frac{\phi}{2N}\right) + \frac{\Delta^2}{2R} \sin^2 \frac{\phi}{2N}$$

In using this method for radii too large to be turned in a lathe, N must be very large, to ensure that ε will be made sufficiently small. Very small curvatures, (i.e., large radii) can be produced entirely by lapping flat plates. When two plates of equal size are lapped, the upper plate tends to become concave, the lower, convex. If the upper plate is the larger, this tendency is enhanced. To maintain cylindricity, the plates cannot be allowed to rotate but can be translated with 2 degrees of freedom. In the case of small radii the large angle ϕ (Fig. 9.21) kept the plate axes always parallel; but for plates of slight curvature, the device of Fig. 9.22 is employed. Fastened to the replaceable oscillating arm, this fixture maintains the plate axes parallel.

Small concave laps of large radius of curvature can be approximated on a vertical milling machine by the use of a fly cutter if the milling axis is

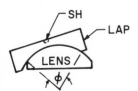

Figure 9.21 Making thick cylindrical lenses.

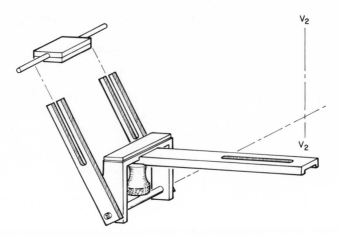

Figure 9.22 An attachment for the two-arm lapping machine (Fig. 9.13) for grinding cylinders of small curvature.

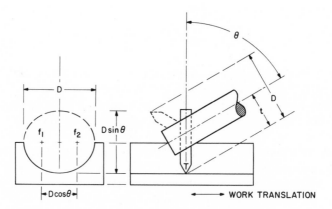

Figure 9.23 Approximating a circular cylindrical lap.

tipped as shown in Fig. 9.23. This produces an elliptical cross section. If the cutting point describes a circle of diameter D, the major axis of the ellipse is D and the minor axis is $D \sin \theta$, where θ is the angle of tip from the vertical. In terms of $t = D/2$, the radius of curvature at the bottom of the cut is

$$R = \frac{t}{\sin \theta}$$

whereas the radius of curvature at the ends of the major axis is

$$r = t \sin^2 \theta$$

At a distance x from the bottom of the curve, the rise is

$$y = t \sin \theta \left\{ 1 - \sqrt{1 - \left(\frac{x}{t}\right)^2} \right\}$$

For a true circle of radius R the rise is

$$y' = R \left\{ 1 - \sqrt{1 - \left(\frac{x}{R}\right)^2} \right\}$$

The elliptical approximation leaves points x too high by the difference between y and y'. This difference can be reduced by recutting with a smaller θ, cutting only toward the edges.

At a point $x = \bar{X}_e$ on the ellipse, the slope is

$$\frac{dy}{dx} = \frac{t\bar{X}_e}{R\sqrt{t^2 - \bar{X}_e^2}}$$

and on the desired circle at a point $X = \bar{X}_c$ it is

$$\frac{dy}{dx} = \frac{\bar{X}_c}{\sqrt{R^2 - \bar{X}_c^2}}$$

These are equal if

$$\bar{X}_e = \frac{Rt\bar{X}_c}{\sqrt{R^2\bar{X}_c^2 + R^2t^2 - t^2\bar{X}_c^2}} \tag{9.1}$$

In this case we have

$$Y_e = \frac{t^2}{R}(1-t)\sqrt{\frac{R^2 - \bar{X}_c^2}{R^2\bar{X}_c^2 + R^2t^2 - t^2X_c^2}} \tag{9.2}$$

whereas

$$\bar{Y}_c = R - \sqrt{R^2 - \bar{X}_c^2} \tag{9.3}$$

Hence if the worktable is lowered by amount $L = \bar{Y}_c - \bar{Y}_e$ as given by Eqs. 9.3 and 9.2 and cuts are made with lateral displacements $\pm H = \bar{X}_c - \bar{X}_e$ with $\bar{X}_e$ given by Eq. 9.1, the result is a closer approximation to the circle. After several such cuts, the approximation can be very close to the ideal.

CYLINDRICAL ENDS ON PLATES

To achieve long path lengths for surface elastic waves, it is desirable to make the path a flattened helix, using both sides and both ends of a crystal plate. This gives the longest path for the smallest crystal. Since the waves will not pass around sharp bends, the ends must be smoothly rounded. It is advantageous to have the ends hemicylindrical with very parallel cylinder axes, paralleled to a fraction of a minute for example.

A complete hemicylinder on a plate end presents problems because a smooth curve cannot be obtained without having the lap edge pass beyond the edge of the area being lapped. If we attempt to get a complete hemicylinder, we must allow the lap edge to strike the flat surfaces, and this damages the surfaces. If we have less than a complete hemicylinder, the curved region does not blend smoothly into the flat region; the surface bends sharply at the join. If the semiangle of the end partial cylinder is I (Fig. 9.24), then for a plate of thickness t the lap radius of curvature must

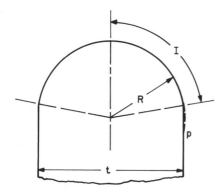

Figure 9.24   Hemicylindrical ends.

be $R = t/2 \sin I$. If $I = 85°$ for $t = 5$ mm, then R must be 2.5096 mm. If R is 10μ smaller than this, the cylindrical surface cannot meet the flat surface. To avoid such tight tolerances we recommend $I = 75°$, which gives a tolerance of 0.176 mm on R. The surfaces join with a 15° angle between cylinder tangent and flat surface. This small area can be later blended in by hand on a flat polishing lap.

If the lap covers the angle $2I$ it must oscillate through an angle less than $2(90° - I)$, otherwise it will strike the flat surface of the crystal and do damage. It seems wise to make the lap angle I' a little smaller than I, permitting the stroke to be increased. The stroke should not exceed $S_{max} = 2\{180° - (I - I')\}$ nor be less than $S_{min} = 2(I - I')$. A safe compromise seems to be $I' = 60°$ which gives $S_{max} = 90°$, $S_{min} = 30°$. With this we set the stroke at about $S = 50°$ (i.e., $25°$ to each side). The lap depth should be $R/2$.

To save wear on the laps we recommend beveling the crystal ends. To keep the cylinder axes parallel to the plate faces and equidistant from the two faces, we need to bevel very carefully, very symmetrically. A vee block with adjustable gap fixture (Fig. 9.25) can serve in this application. A bit of the crystal edge projects through the gap. The crystal is

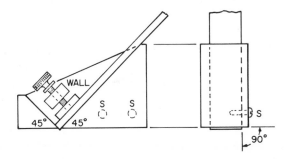

Figure 9.25 Fixture for beveling crystal ends.

clamped in place by means of the knurled screw, and the edge is ground flush with the fixture bottom. When two screws S, S are loosened, the right-hand block can be slid to adjust the gap. The fixture is placed on a flat surface and a cylinder gauge is placed in the vee. The gap is closed until the cylinder touches the flat surface and the two walls of the vee simultaneously. The screws S, S are then tightened. The choice of cylinder depends on the crystal thickness t, the angle ϕ and the thickness Δ of any protective tape that covers the crystal faces. The diameter of this cylinder gauge should be

$$D = 2R + \left\{\Delta - t\left(\frac{\csc \phi - 1}{2}\right)\right\} 3.414$$

For $\phi = 75°$ this gives us $R = 0.5176t$, $D = 0.975t + 3.414\Delta$.

The small laps can be made by the methods suggested by Figs. 9.18, 9.19, and 9.20, but the polishing tool should be polishing pitch. In Fig. 9.20 ϕ should be given the value of I' arrived at earlier—namely, $72°$. The pitch lap for small radii presents a special problem. A trough (Fig. 9.26) can have its ends sealed with masking tape, then heated and filled with pitch. After cooling, the tape is removed. The pressing tool has a foot f that keeps the presser plate p upright in use. The plate p has a lower edge made hemicylindrical with the same radius used for the lap, and it can be

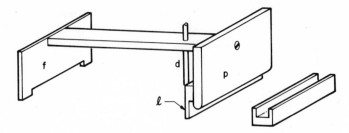

Figure 9.26   Pitch-pressing tool for polishing hemicylindrical ends.

ground by such a lap after being milled with a radius cutter. A ledge plate
ℓ places the presser at the center of the pitch-filled polisher trough if the
trough is pressed against ℓ. A depth stop d is adjusted so that the presser
cannot sink into the trough to a depth greater than is allowed by the
motion of the polisher in use. Since the face of p is left free of projections
such as screw heads, it can be placed on a hot plate and heated up quickly.
The trough sits on the table, full of pitch and covered with a Mylar sheet.
A drop of water on the pitch helps to hold the Mylar in place. The warm
presser is then placed over it with the trough against the ledge and the foot
f resting on the table. A weight is placed over p and the assembly is
allowed to cool. If f is properly adjusted, the presser will enter the pitch to
the same depth at both ends, and this depth will be proper if d is set
properly.

   The device of Fig. 9.22 is not suitable for making hemicylindrical ends
on long plates because the length of plate cannot be accommodated by the
device. If the plate width w is quite large as compared with the plate
thickness, laps can be driven by the usual simple drive pin as in Fig. 9.27.
The width w can be the combined width of several plates secured

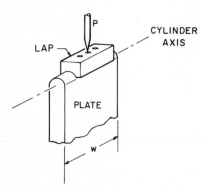

Figure 9.27   Lapping cylindrical ends on a
wide plate.

together. Normally pin P (Fig. 9.27) is placed in the central pit of the three pits on top of the lap. If one edge must be ground more than the other edge to make the cylinder axis perpendicular to the plate edges, the pin is placed in the hole nearest the high edge. The stroke is then readjusted and the grinding proceeds until the cylinder axis is perpendicular to the plate edges. The polisher is used last, to finish the plate end.

If a single narrow plate must have its ends lapped round and polished, one can either use dummies to make a composite of great enough width w or try to process the narrow plate alone. The simple pin drive is troublesome when a narrow plate is being processed alone. The tool should be shorter than the width w, and if the drive pin is much above the cylinder top, there is excessive torque, tending to turn the lap over about a horizontal axis perpendicular to the cylinder axis. Torque can be reduced by making the laps very thin, but then the action tending to oscillate the lap about the cylinder axis becomes overly sensitive.

Driving the lap along the cylinder axis requires a thin lap; oscillating it about the cylinder axis requires a thick lap. We can effectively have it both ways if we use the gimbal mount of Fig. 9.28, where L is the lap and C is a lap carrier fastened to the lap by a screw or screws. There is a center hole at each end of C, engaged by pivot screws PS, which have 60° points. The pivot screws are screwed into the gimbal ring GR, the threaded holes being split as shown, and squeezed together enough to ensure that the screws are tightly held and will not loosen in use. On each side of the gimbal ring there are three pivot holes. Normally the two pivot screws in the frame F engage the two center pivot holes. However if it is

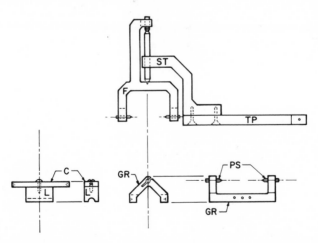

Figure 9.28   A lap-driving device good for end lapping of thick narrow plates.

necessary to grind one end of the hemicylinder more heavily than the other end to improve squareness, two of the outer pivot holes are used. The frame F is in turn pivoted on a stanchion ST. This is carried by a tee plate TP. Pivot holes PH at the ends of the crossbar of the tee engage pivot screws carried by a sliding frame that is carried by one of the oscillating arms. The sliding frame allows the lap to be centered over its work, and the pivots allow the lap assembly to be swung up from its work. Squareness of hemicylinder with plate edge is checked by means of a laser setup described in Chapter 12.

10

The Making of Complete Spheres

LARGE SPHERES

The most economical procedure for making large spheres is to start by cubing the material. It does not matter if the corners and edges are missing because our next step is to saw off the eight corners, as in Fig. 10.1. To do this symmetrically, we place the cube (of dimensions

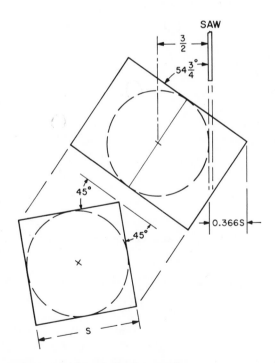

Figure 10.1   Preparing a cube for grinding into a sphere.

251

$S \times S \times S$) in a vee block, turn the vee block $54\frac{3}{4}°$ from the plane of the saw, and saw off an amount $0.366S$. This makes a surface tangent to a sphere of diameter S imagined to be inside the cube. When the eight corners have been sawed off, the block should look like the illustration in Fig. 10.2. The greatest dimension should now be about $1.23S$, whereas in

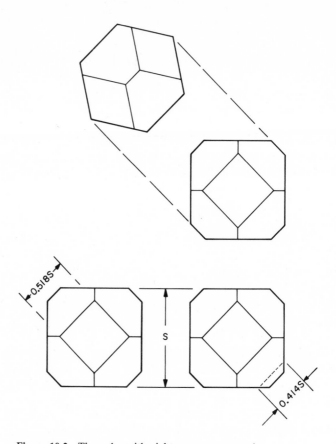

Figure 10.2 The cube with eight corners removed.

the cube it was $\sqrt{3}S = 1.732S$. We can now either grind off slightly the many sharp points and edges or proceed directly to the next step. To grind off the sharp points we merely hold the piece over a rotating abrasive-loaded lap to round off the piece as judged by eye and grind to make faces $0.414S$ wide as indicated in Fig. 10.2. With smaller spheres we can proceed directly to the "two-tube method" of sphere making.

THE TWO-TUBE METHOD

The two-tube method is illustrated in Fig. 10.3. The roughed out piece is placed between the ends of two tubes whose outer diameter is slightly larger than that specified for the finished sphere. The inner diameter is about $\frac{7}{10}$ as large as the finished sphere diameter. The ends are beveled at about 45° as shown. With wet abrasive applied to the "embryo sphere," the

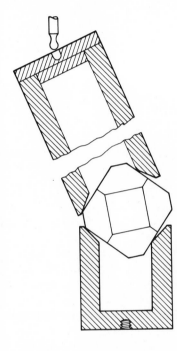

Figure 10.3 Grinding the sphere.

lower tube is rotated about its axis, the upper is swung back and forth. This can be done with the optical engine shown in Fig. 8.33. The sphere should keep shifting around in its seat, giving all parts equal chances at the grinding action of "sphere sliding over tube." Watch carefully that the "sphere" does not become wedged into one tube or the other and be ground nonspherically. Less care is needed as the "sphere" becomes "rounder," since nature causes the piece to keep shifting in its seat, ensuring that grinding is equal all around. Quite accurate spheres can be made by this method.

THE RUBBER DISK METHOD

Sawing off the corners is impractical for somewhat smaller spheres, and we often choose to remove the corners by rolling the crystal cube between two elastomer disks. One disk is stationary, the other rotating. The cube is retained in a special washer, and the grinding action is produced by the crystal corners wearing on the bore of the washer. The upper disk is rotated by the drive shaft of a slow-speed drill press, as in Fig. 10.4, where

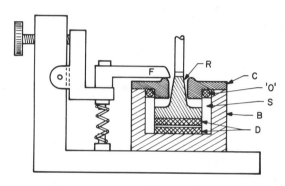

Figure 10.4   Rough grinding small spheres.

D, D are the elastomer disks (generally neoprene), B is the housing body (plastic), "O" is an O-ring, C is the cover (plastic), R is the rotor (plastic), and S is a readily removable, abrasion-resistant sleeve. The sleeve is made of metal, glass, or ceramic. A finger F forces the cover against the O-ring sealing the unit. Centrifugal force "seals" the central hole of the cover against escape of the fluid vehicle of the abrasive.

In Fig. 10.5 the retainer "washer" appears in more detail: one edge of the cube is in contact with the lower rubber, the upper rubber is in contact with the opposite edge of the cube. As the upper rubber turns clockwise, it rolls the cube and drags the retainer, thus rubbing on the forward edge of the retainer ring. Centrifugal force causes the retainer to contact the sleeve and roll about inside the sleeve. This allows retainer to wear evenly all around. The cube shifts axes occasionally unless the dimensions are too unequal. A square prism that is longer than its width and thickness tends to roll mostly about its long axis. To eliminate this condition, we let the piece run for a short time without a retainer, permitting the long end to wear away on the sleeve. When equidimensionality is attained, the piece can be placed in a retainer and ground.

The hole in the retainer should be about 1.75 the length of the side of

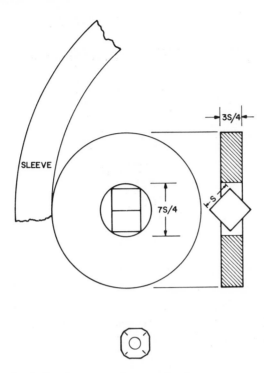

Figure 10.5   Rough grinding the corners from small cubes.

the cube, the ring thickness should be about three-quarters the length of a side of the cube. After some minutes of grinding the corners are sufficiently reduced that the piece can be ground in a new retainer with a hole of diameter about 1.6 times the side of the cube. Sometimes it is worth finishing this phase by grinding in a retainer with a hole of diameter about 1.3 times the side of the cube. The final product of this first roughing operation is illustrated in Fig. 10.5 (lower inset). The dimensional ratio, that is, the ratio of maximum diameter to the minimum diameter, has been reduced from about 1.73 (as it was in the cube) to about 1.4.

The dimensions and dimensional ratio are checked occasionally under a microscope with a micrometer eyepiece. The sphere is placed on a smooth, horizontal surface, and gravity causes the smallest dimension to be vertical. The sphere is then picked up with special tweezers held horizontally. Having a square shank, the tweezers will remain as placed on any of the four sides. The tweezers (Fig. 10.6) are normally closed but can be opened by pressing on the areas P, P. The jaws close to grip the sphere, which can then be measured in three mutually perpendicular

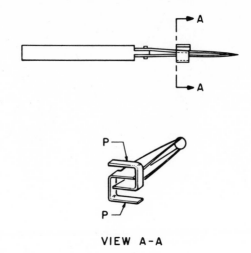

VIEW A-A

Figure 10.6   Tweezers for handling small spheres.

directions by placing the tweezers under the microscope in three different positions, as in Fig. 10.7. The dimensional ratio is taken as the greatest diameter divided by the least diameter.

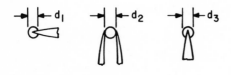

Figure 10.7   Measuring small spheres held in tweezers.

SMALL SPHERES

Probably the most economical method of generating spheres from small random pieces is the pneumatic method. The original idea is depicted in Fig. 10.8. Here the small pieces that are to be ground to spherical shape are placed in the cylindrical cavity whose cylindrical wall is covered with an abrasive coating. Air enters the cavity nearly tangentially, swirls around a few times, and escapes through the central exhaust port that is covered with a screen to prevent the workpieces from being blown out of the cavity. The air stream forces the crystals to roll around the cylindrical wall, and in so doing the crystals frequently enter the jet stream near the air entry point—where the air velocity is a maximum. A perfect sphere entering this jet stream receives mainly a translational impulse in the direction of the jet air flow. If there is a lump on the sphere, however, the

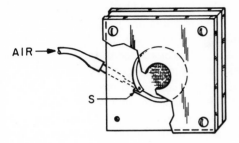

Figure 10.8 Pneumatic grinding of small spheres.

lump receives its own impulse, and the result is a spin of the "sphere." This causes the "sphere" to strike the abrasive surface, spinning in a way that serves to reduce the lump. In a sense, a prolate sphere has two such lumps, and the spinning action tends to reduce the prolateness by grinding away these two lumps.

The grinding action is greatest at the first impact of the sphere against the abrasive surface, and the grinding surface is soon deeply eroded at this point. A design that uses this effect appears in Fig. 10.9. An abrasive stone is clamped into place near the jet and takes the brunt of the wear. It can be turned over to present a new face after excessive wear, and the surfaces can be reground flat and finer stones can be used after the rough grinding. The design was used with liquid propulsion in grinding some crystals that could not be taken out of the growing solution nor allowed to come into contact with metals. A piston pump made of plastics and using O-ring obturation squirted the liquid into the circular race where the sphere was ground on the stone. The liquid was returned to the pump. However liquid propulsion is greatly inferior to air propulsion.

Spheres slightly out of round may be turned by forcing them to rotate in

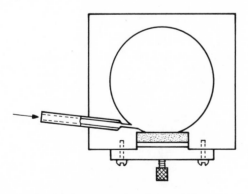

Figure 10.9 Second version of the pneumatic sphere grinder.

a tube end. (See section "The Two-Tube Method".) The two-tube method sometimes fails because the ball becomes stuck in one of the tubes and is quickly ruined. If the sphere is placed on a smooth rubber mat, then caught in the conical hole in a tube end, it can be forced to roll by moving the tube about over the mat. An abrasive slurry now grinds the ball into the cone. If the ball is fairly far out of true, say 5%, it may occasionally ride out of the hole. If a dry mat were used, the ball might fly some distance and be lost. However the fluid prevents this. Such rejection is also discouraged by having the cone angle small and the ball rather deep in the hole, as illustrated in Fig. 10.10. The seat must wear away as well as

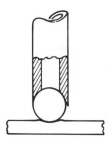

Figure 10.10   Lapping a sphere in a tube.

the ball, and if the ball sinks in so that it no longer presses on the mat, the tube end must be cut back. The hole in the tube is important because metals grind away much more slowly than brittle crystals. This means that the ball will shrink faster than it can recede into the lap seat, producing a situation represented in Fig. 10.11.

If the ball is really a prolate spheroid and the tube is driven with a circular motion, the ball will keep swinging around in its seat so as always

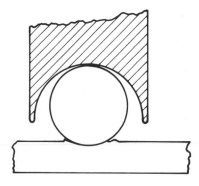

Figure 10.11   Illustration of the importance in sphere making of a central hole in the lap.

to roll around its prolate axis—it "steers" itself. It can soon resemble a lemon. This undesirable feature can be avoided by using a square stroke, which forces the prolate axis to yield because of the 90° changes in direction.

The machine in Fig. 10.12 is designed to give the above-mentioned square stroke, or rather, a satisfactory substitute for it. The machine is built about a square, and two opposite corners are "pivot points" for

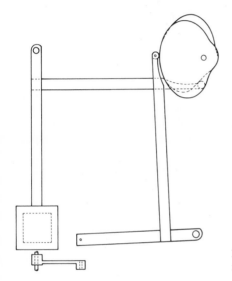

Figure 10.12 Square stroke sphere-lapping machine.

swinging arms. The upper left axis allows an arm carrying the mat to oscillate through about 10°. The lower right axis allows an arm supporting the lap to oscillate about 10°. These oscillations are forced by two cams, one above the other. Springs (not shown) force rollers against the cams, which are driven by a small-geared motor mounted below.

The cams are designed to give a sinusoidal motion, but there is a momentary pause at each corner. This prevents slight errors in the cams from causing the direction of motion to change before the old motion is completed. This would result in a rounding of the path corners and a chance for the ball to steer itself around the corners.

In Fig. 10.12 the bracket that supports the lap is removed and appears just below the mat. The ball track on the mat is indicated by the dotted line. This machine produces a highly precise sphere.

PARTIAL SPHERES

Partial spheres are most easily obtained by grinding them on a metal lap of the proper curvature. However sometimes it is preferable to cement several pieces of crystal together, produce a sphere from the assembly, and finally separate the pieces.

Templates

For partial spheres of more than a few millimeters size it is wise to first prepare templates of the required curvature. These show us when lap and lens have the proper curvature. If the curvature of the surface does not match the curvature of the template, it is obvious which way to go in correcting the surface—shallower or deeper. A template also reveals departures from uniformity of curvature—something a spherometer cannot do as well. Commercially available radius gauges (templates), commonly used in machine shops, are somewhat limited in range. Odd size gauges must be specially made. Measurement under the microscope is preferred for very small partial spheres (see Chapter 12 on measurement of curvature).

Templates are easily made of sheet metal if a lathe is available. In Fig. 10.13 two slips of sheet metal M, M' are soft soldered to the face of a brass bar at the ends or secured with screws. The bar is mounted in a

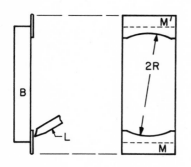

(a)

Figure 10.13   Making templates for a convex lens.

four-jaw chuck in the lathe, and the inside ends are cut with the lathe tool L until an inside caliper shows the proper value for $2R$. The outer ends could be cut as well, and an outside caliper used to establish the proper value of the radius of curvature. After unsoldering and removing the solder, an accurate pair of templates remains. If only one metal tab is soldered to the bar, it is more difficult to measure the radius. For small convex templates it is easy to put a sheet metal disk on an arbor and turn the periphery to the proper radius, as in Fig. 10.14.

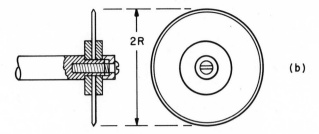

Figure 10.14   Making a template for a concave lens.

Making the Laps

We can approximate the spherical surface in cones as diagrammed in Fig. 10.15. For a two-cone approximation we cut into the end of a squared metal cylinder at an angle $3\theta/4$, where $\sin \theta = r/R$, as in Fig. 10.15a, the width of the cut being $2R \sin \theta/4$. We then reset the compound rest to the angle $\theta/4$ and remove the central area to the center, as in Fig. 10.15b. For very deep curves we can use a three-cone approximation (Fig. 10.16), in which the outer cut is made at an angle $5\theta/6$, where $\sin \theta = r/R$, the next cut is made at an angle $\theta/2$, and the inner cut is made at an angle $\theta/6$. Each portion is of width $2R \sin(\theta/6)$.

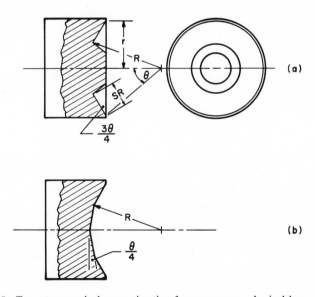

Figure 10.15   Two-stage conical approximation for a concave spherical lap.

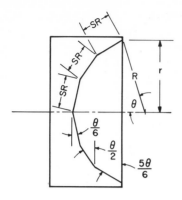

Figure 10.16   Three-stage approximation for a concave spherical lap.

The illustrations have shown only concave laps. However, convex laps are readily made using the same angles and dimensions. We can either reverse the signs of all angles or keep the angle signs the same but run the lathe backward and work the tool on the far side of the lap.

Once a lap has been made by such an approximation method, the surface can be scraped to achieve a fit to the template (see Fig. 10.17). A file whose end is ground rounded is rested on a tool holder in the lathe, and as the work revolves material can be scraped from the high areas. Frequent checking with the template helps in deciding what regions need attention.

Since we cannot make a template for laps of very small curvature, we make a concave and a convex lap by the cone approximation method. We

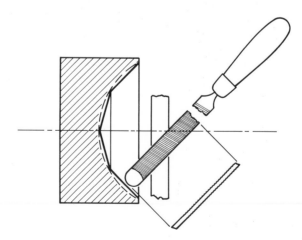

Figure 10.17   Scraping the lap to improve sphericity.

then grind the two to fit, as in Fig. 10.18. The lower lap is rotated about its axis, the upper lap is held with its center in a fixed position by a ball-ended pin P. The upper lap is free to rotate about this pin and does so. The pin is set off the axis of the lower lap at a distance such that the high ridges of the convex lap grind into the high flats of the concave lap. A slight oscillation of pin P about its central position is good, but the pin should

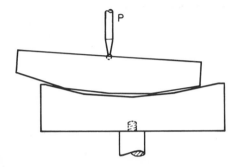

Figure 10.18 Grinding two approximations to improve sphericity.

never approach the axis of the lower lap because the two will lock together and it will be difficult to restore the pin to its proper position. The curvature can be checked occasionally with a spherometer (see Chapter 12).

Since in grinding two laps together as in Fig. 10.18 the upper lap tends to become more concave and the lower to become more convex, it is well to interchange positions periodically to maintain the proper curvature. One can start with flat laps and use this effect to produce very slight curvature. The greater the overhang, the faster the change of curvature. If the process is speeded up too much by excessive overhang, however, the surfaces cease to be spherical. An overhang of one-sixth of the diameter is the limit, if sphericity is to be maintained. Early stages in which the one-sixth overhang rule is exceeded can be corrected by later grinding with greatly reduced overhang.

A great deal of material may need to be removed in making a lens of short radius. Such extensive use may alter greatly the curvature of the lap used for roughing. In some cases it is advisable to make two laps, one for roughing, the other for finishing. If a great many such surfaces are to be ground, it may pay to procure a spherical generator machine.

The Spherical Generator

In Fig. 10.19 a lens L is being made by grinding with the diamond cup wheel DCW. Lens L rotates slowly about an axis AA, which makes an

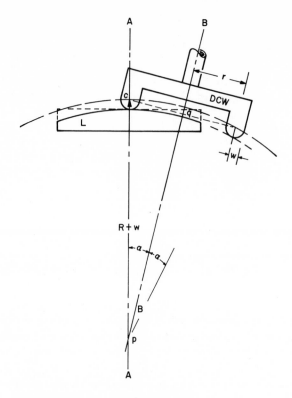

Figure 10.19   The spherical generator.

angle α with the cup wheel axis BB. Axis BB swings about a point c at the center of the curved cutting surface of the cutting wheel; thus α can be given any value between $-90°$ and $+90°$. The cutting surface of the wheel has a radius w. Axes AA and BB intersect at point P and from right triangle cpq we see that

$$\sin \alpha = \frac{r}{R + w}$$

If α is made negative, we obtain a concave surface. Commonly w is negligible compared with R. It is readily seen that quite large radii can be made by this method. If $r = 5$ cm and $\alpha = 1°$, then $R + w = 286.5$ cm.

Making Laps

The spherical generator method can be used to make laps of large radius, as well. A fly cutter on a vertical milling machine sweeps out a circle of radius R. If the axis is tilted to angle α we have

$$R = \frac{r}{\sin \alpha} - w$$

The workpiece must be mounted on a swivel table or index head set for zero angle and accurately centered and rotated through 360° many times. The worktable is raised slightly after each rotation of the work. If the center of the lens L is directly under point c (Fig. 10.19), it will remain there as the table is raised. It is interesting to note that although this method was used to approximate circular cylinders in the previous chapter, it really produced elliptical cylinders. However here we obtain true spheres: while the cylinder has been slid along under the tool, the sphere has been rotated under the tool, and this makes the difference. Milling of spherical laps is both easier and more accurate than the multiple-cone method of approximation. It is also possible to mount the lap blank in a chuck on an indexing head and tip the head to the angle α. This allows us to leave the vertical Miller axis set at $\alpha = 0$, which is advantageous because after this setting has been disturbed, it must be very carefully "zeroed in" again to mill flat surfaces. The tipped index head presents one disadvantage. With the cutter axis tippable, the work center stays under the point c (Fig. 10.19) as the table is raised; but with the tipped index head the center c shifts relative to the work center. If care is taken to make the last cuts with the tool passing through the work center, a proper sphere is produced; otherwise a more complicated surface results. More is said on this subject in Chapter 11.

Gimbals for Thick Lenses

Although plates can be successfully lapped by the simple technique given in Fig. 10.20a, such is not the case for thick plates or for rods requiring curved surfaces on the ends. In Fig. 10.20a as the lap L revolves, the ballpoint pin BPP (engaged in a shallow hole in the mounting plate P) forces the mounting plate to move across the lap, carrying the work with it. The arrow OK shows the point of application of this force. If the plate is thick, as in Fig. 10.20b, the driving point is very close to the center of the curvature, hence cannot "drive." We can lower the driving point by means of a gimbal as in Fig. 10.20c, where Y is a yoke holding the crystal and G is the gimbal ring. The gimbal ring is driven by a fork F from two points 90° around the ring from the pivot point screws PS (Fig. 10.20d). When it is permissible to mount the work directly on the turntable of the machine, the lap can be worked as in Fig. 10.20a, dispensing with the complications of the gimbal ring. With the gimble rig

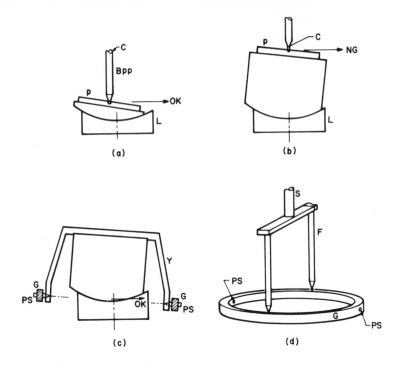

Figure 10.20   Problems with thick lenses.

fork, however, the shaft S can turn in a bearing, allowing the work to revolve freely as it is ground by the lap.

Figure 10.21 presents a simpler method for use with long slender rods terminated by deep curves. The stability of the arrangement depends on the curve being deep. For very shallow curves the rods could be supported by outlying feet and lapped as in Fig. 10.20a. In Fig. 10.21 the rod R is lapped by the lap L. The rod is extended by means of an extension rod ER, which terminates in a small shallow hole that takes the ball of a ball-point pin BPP. The ball-point pin is carried on the sweep arm of the optical machine (Fig. 8.33). With this device we can make either concave or convex ends on rods. Starting the pit for a concave end is easily done with the guide cap GC (see inset, Fig. 10.21). When the pit is deep enough to make the assembly stable without the guide cap, the cap is best removed.

A useful variation of the machine just described appears in Fig. 10.22. Here both the lap and the tool are driven with a common belt, but the angle does not change by oscillation. Since both components are driven

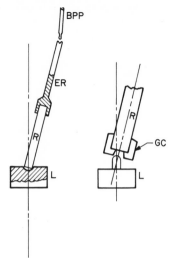

Figure 10.21 Grinding very small radii on thick lenses.

and the inclined member is floating between an upper pivot and the grinding area, the lap and the tool mutually grind each other, generating spherical shapes. The belt on the inclined member pulls at right angles to the ball-pin bearing, hence does not pull the inclination to either greater or lesser angles. It readily produces quite accurate lenses very quickly. For concave lenses the ball tool can be prepared by filing by eye, knowing that mutual grinding will improve the curves.

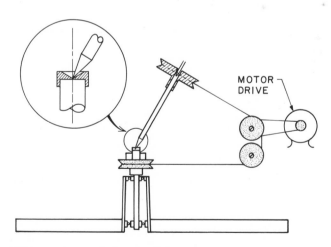

Figure 10.22 Variation of grinding machine in Fig. 10.21.

In Chapter 9 we demonstrated that a rod that measures the same diameter all around need not be a circular cylinder. The same is true of spheres. If a pseudo-circle is constructed about an isosceles triangle as shown in Fig. 9.10, and the axis of symmetry is used as a rotation axis, a pseudo-sphere results. All diameters as measured with parallel-jaw calipers are the same, but the body is not a sphere because there is no one point within the body that is equidistant from all points on the surface. In a manner similar to that reviewed in Fig. 9.10, we can construct pseudo-spheres based on polyhedra, which are not surfaces of revolution. Hence we emphasize that if all the diameters of a body are equal, the body is still not necessarily a sphere.

11

Making Doubly Curved Surfaces

DOUBLY CURVED SURFACES

A doubly curved surface is a surface that has, at any point, different curvatures in different directions. Examples are cylinders, ellipsoids, toruses, toroids, and paraboloids. Much more complicated surfaces can be made in lathes and milling machines by moving work and tool, each in predetermined ways, but this approach does not produce surfaces of optical quality for two reasons. The first reason is machine crudity: work supports and tool supports are elastic, and the work of cutting can cause forced vibrations. When such vibrations are obvious to the unaided eye, they are called tool chatter. Also bearings are nonideal. Ball bearings may support a shaft that revolves, and the instantaneous axis of rotation may wobble around inside a circle of about a micron's diameter. But optical precision sometimes demands accuracy better than 0.050 μ. Sleeve bearings are no better. The second reason is the tool marks left by cutting tools. These marks are local departures from the ideal surface, and they are larger than 0.050 μ.

Lapping and polishing are done by form-fitting tools that follow the surface being worked. The tool and the work wear each other away to achieve this fit—aided by abrasive, which does the cutting. In polishing the abrasive may be of the order of a fractional micron in diameter, and most of this diameter is buried in the tool; hence "tool marks" are very small. The abrasive-loaded tool must pass over every point on the work many times and along more than one path. Since the tool "floats" on the work surface, rigidity of driving rods and perfection of bearings is of little importance. What is important is the requirement that the tool fit the surface everywhere it is driven; in addition, as many points on the tool as possible must pass over every point on the work surface and along more than one path, preferably along all paths through the point. Because of this spheres and cylinders can readily be lapped and polished, but other shapes present problems calling for compromises.

269

For example, two partial spherical surfaces that match (i.e., have equal curvatures of opposite sign) can be fitted together perfectly. They touch each other at all points of the region of overlap. One surface can be slid over the other in any direction without spoiling the fit. Either one can be rotated relative to the other by any amount without spoiling the fit. If of two approximately matching surfaces one is a lap or polishing tool and the other is a workpiece, each point on the workpiece can be passed over in many directions by many points of the tool. The work and the tool wear each other away, and when they have ground each other until they fit in all positions, both must be partial spheres.

When fitted together, two matching partial circular cylinders can be slid over each other in any direction. However one cannot be rotated relative to the other so that their axes are no longer parallel. Hence there is restriction of motion, but enough degrees of freedom exist to make possible a useful product.

In principle, if mathematically ideal surfaces were required, only spheres and cylinders could be made by lapping. Here we ignore molecular dimensions as compared to one-tenth wavelength of light. That is we ignore, say 10 Å as compared with $5000/10 = 500$ Å. As a practical matter we now examine the end use of the surface to be made, to determine whether we can relax our standards a bit. If the surface is to reflect light to form a sharp image, we can tolerate errors of perhaps one-eighth wavelength of the light. Hence we can make other doubly curved surfaces slide small distances over each other in forbidden directions. This permits the using of surfaces of revolution—toroids, for example. A toroid is the surface generated by any plane closed curve rotating about a straight line in its own plane. Matching toroidal surfaces can be slid over each other freely in the direction of the generating rotation, but if they are slid over each other more than slightly along paths normal to this direction, excessive error will result. They cannot be rotated relative to one another.

THE TORUS

If the generating curve is a conic section, the generated surface is called a torus. Most commonly the generating curve is a circle, and we now examine this kind of torus. We are mainly interested in only a part of the half-torus in Fig. 11.1—namely, the part below the line CC. This part has a radius of curvature R in the plane of the paper and a radius of curvature r normal to the paper. The small circle of radius r was swept around the axis AA, which is at a distance $R - r$ from the center of the small circle.

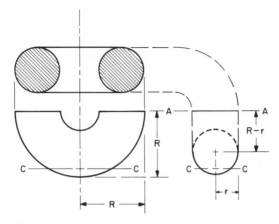

Figure 11.1 Generating a simple torus (concave).

Figure 11.2 illustrates a second way in which the two radii R and r can define a partial torus. Where the larger circle of radius R is turned about an axis $A - A$ distant $(R - r)$ from the center of the large circle. Again only the half-torus is shown, and again our interest is confined to the small part below the line CC. We generally prefer the first kind of torus of principal radii R and r, since the direction of free motion is along the length and with the larger radius of motion. Faster grinding and polishing thus are possible.

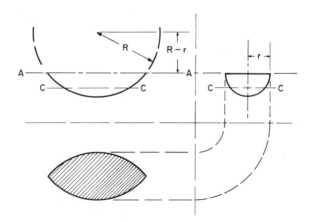

Figure 11.2 Generating a less useful torus (concave).

Making Toroidal Laps

With a vertical milling machine and an indexing head, we can generate not only the tori just mentioned and the elliptical cylinders discussed earlier, we can make more general toroids. Let us start with toric laps, which are of course a special case of toroidal laps.

The Toric Lap

Figure 11.3 presents a method of making toric laps. The lap block LB is secured to a mounting block MB, which is in turn fastened to a mounting disk MD. The mounting disk is held in the chuck of an indexing head and can be rotated about the index head axis. The cutting tool CT can be positioned so that the cutting point just touches the center of the lap block. The cutting tool is then spun around its axis, the index head is rotated a little and the cross-feed advanced, so that as the index head is rotated back, a cut is taken across the lap block. The cross-feed is again advanced and a new cut is taken across the lap block. This is continued until the depth is sufficient to give the desired value of R. The work is checked periodically by shutting off the power, turning the tool away

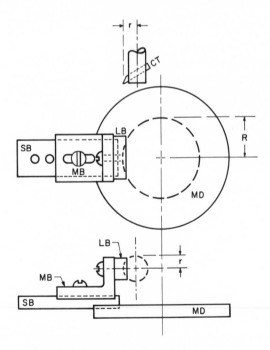

Figure 11.3 Actual machining of a toric lap (concave).

from the lap, and measuring the distance from tool point to bottom of depression—it should be $2R$.

More General Toroids

Our simple machinery—vertical milling machine and indexing head—allows us to make only surfaces that can be generated by rotating a circle about an axis, but the axis does not need to be in the plane of the circle made by the rotating tool point. This restriction may seem to exclude the surface so generated from the class of toroidal surfaces. We can deduce the equation of a curve which is the cross section of the generated surface, however, and this permits us to assume that the surface was generated by rotating this curve about an axis in its own plane. This fills the requirements of a toroid.

The conditions for making an approximately elliptical toroid with the major axis of the "ellipse" parallel to the generating axis are indicated in Fig. 11.4. The cutting tool is below the index head axis by amount l and is

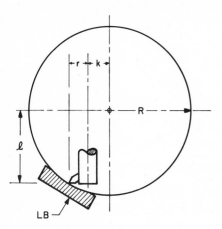

Figure 11.4 Machining a less simple toroidal surface (concave).

off center by amount k, horizontally. Again the cutting point sweeps a circle of radius r, and R is given by

$$R^2 = (k + r)^2 + l^2$$

We can show algebraically that the generating curve is

$$y^2 = x^2 \pm 2k(r - \sqrt{r^2 - x^2})$$

Its shape appears in inset (Fig. 11.5). The radius of curvature at the bottom of the lap (point b) is $Rr/(k + r)$. An approximate equivalent to

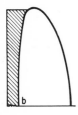

Figure 11.5   Cross section of the toroidal surface generated in Fig. 11.4.

the useful part of the curve is the ellipse of major axis $2r$, minor axis

$$2\{R - \sqrt{k^2 + l^2}\}$$

DOUBLY CONVEX LAPS

The doubly curved laps just described were all doubly concave. Doubly convex laps can be cut on the same combination of indexing head and vertical milling machine. One such operation is illustrated in Fig. 11.6.

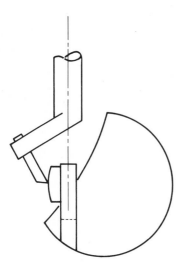

Figure 11.6   Machining a convex toroidal lap.

This arrangement can be varied by cutting below center or by tilting the tool axis. Doubly convex laps can also be cut in a lathe by cutting and filing the edge of a disk to fit a template. When the edge fits, laps are cut from the disk.

Doubly curved convex-concave laps can also be made on the milling machine (see Fig. 11.7). These saddle-shaped laps can also be turned in a

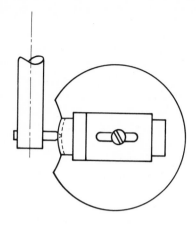

Figure 11.7   Machining a saddle-shaped lap.

lathe as guided by a template, the final lap being cut from the disk when a proper fit is achieved.

LAPPING SURFACES OF DOUBLE CURVATURE

Having prepared laps of appropriate shape, we now wish to use them to make crystal plates with surfaces of double curvature.

For simple work with curves that are not too deep, we can use the optical machine of Fig. 8.33 fitted with the attachment in Fig. 11.8. The hollow rectangular frame screws onto the turntable of the optical

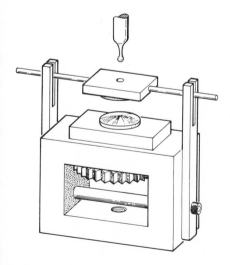

Figure 11.8   Grinding a bent axis mirror.

machine. The notched wheel inside the frame is the head of a machine screw that can engage a tapped hole in the lapping block or work holder. The slotted bars at the ends of the frame are rigidly connected by means of the shaft that passes through the frame. Two bars are screwed into the workpiece mounting block (or tool). These engage the slots of the slotted bar and prevent relative rotation of tool and workpiece but allow translation in any direction. As the rectangular frame is rotated by the optical machine, the ball-point rod (Fig. 11.8, top) oscillates back and forth, thus sliding the work over the lap.

If the work in the direction of small curvature is longer than it is in the transverse direction, as often happens, having the strokes of equal length in all directions is of no great importance. For more exacting work, however, separate control over strokes is necessary along the length and across the width.

A simple machine that gives separate control over length strokes and cross strokes is presented in Fig. 11.9. Here a drive wheel D1 is mounted on the shaft of a geared motor (not shown) under the base plate BP. An adjustable eccentric in D1 causes the lower work holder mounted on the mounting plate MP to oscillate with a desired amplitude. Drive wheel D1 also has a vee groove for a belt that drives D2, a second drive wheel that causes the upper work holder to oscillate with the desired amplitude. Both D1 and D2 are set for zero amplitude, but D2 is shown from two other angles in the detail above Fig. 11.9. D1 and D2 are similar, but since D1 is

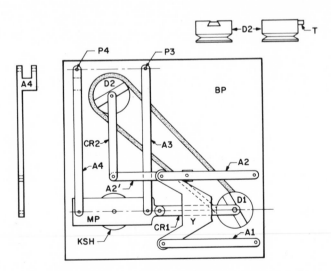

Figure 11.9 Lapping machine for deep, doubly curved surfaces.

longer and has a deeper vee groove, the oscillation rates are distinctly different. Also D2 is adjustable by loosening a top screw, whereas in D1 the tongue T is locked by means of a side screw because the top is covered with other elements of the machine.

The arms A3, A4 oscillate about the stationary pins P3, P4, and hence carry the mounting plate MP through a short arc that is straight enough for our purposes. (A side view of arm A4 appears at the left in Fig. 11.9.) The mounting plate length direction remains parallel to two edges of the base plate BP as the mounting plate oscillates. The right end of the mounting plate is fastened to the connecting rod CR1, and rotation of D1 causes oscillation of MP. The knurled screwhead KSH can be turned with the fingers to screw down the workpiece.

As drive wheel D2 turns, connecting rod CR2 causes arms A1 and A2 to oscillate, A2 being connected to CR2 by means of the extension rod A2′. Arms A2 and A1 are connected to the yoke Y, which consequently oscillates through a short arc, the yoke axis always remaining parallel to two edges of the base plate. For clarity the upper work holder system is shown separately in Fig. 11.10.

In Fig. 11.10 the yoke Y carries a slide S, which can be slid along the tongue of Y and locked where desired. The slide S has a dovetail groove that takes a cross slide CS that can be locked in any desired position in the

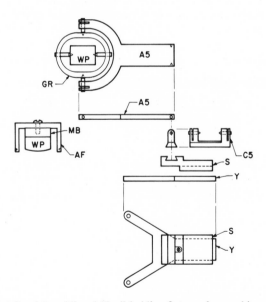

Figure 11.10   Details of "work" and "lap" holding fixtures for machine of Fig. 11.9.

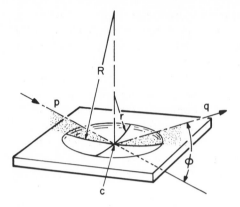

Figure 11.11   Relation between surface
curvatures, object distance, and image
distance for bent axis mirrors.

groove of the slide S. The cross slide CS carries two pivot screws that
enter pivot holes of arm A5. The pivot screws can be adjusted by means
of a hex wrench. The threaded holes for the pivot screws are slotted, and
part is sprung to keep the slack out of the screws. At its left end, arm A5
has two pivot screws that engage pivot holes in the workpiece WP,
although for thick convex pieces the auxiliary frame AF is used. This
places the pivot holes almost in the plane of the grinding resistance
forces.

With this machine it is possible to adjust long and cross stroke lengths
independently and to center or decenter either stroke at will. This
refinement is essential in making double curvature pieces to specification.

It should be pointed out that mirrors with such double curvatures will
focus an image at an angle, that is, "bend the axis." To bend the axis by
angle φ we must have the "in plane" radius R:

$$R = \frac{2f}{\sin \varphi / 2}$$

and the across radius r:

$$r = \frac{2f \sin \varphi}{2}$$

Here of course the object distance p and the image distance q are related
through

$$\frac{1}{p} + \frac{1}{q} = \frac{1}{f}$$

These conditions are pictured in Fig. 11.11.

12

The Measurement of Curvature and Prism Angles

MEASUREMENT OF CURVATURE

The easiest way of determining the curvature of complete spheres is to measure the diameter with calipers. The curvature is then the reciprocal of the radius. Even large spheres can be measured by methods that measure the local curvature, however. The principal tool for this is the spherometer, which measures curvature using only part of the surface for the measurement—it is most helpful on partial spheres but can be employed on cylinders or surfaces of double curvature.

The Spherometer

The device in Fig. 12.1 is a very practical spherometer for general use. A dial gauge is used with any one of several attachments. The attachments slide onto the sleeve of the gauge so firmly that the assembly feels

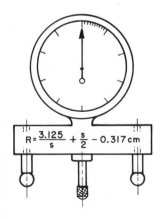

$$R = \frac{3.125}{s} + \frac{s}{2} - 0.317 \, cm$$

Figure 12.1 A spherometer.

solid—no shake of the sleeve. In a plane that includes the axis of the sleeve are the axes of two "legs." These legs are equally distant from the sleeve axes, their axes being parallel to the sleeve axis. The legs terminate in steel bearing balls soldered into conical depressions at the ends of the legs. The accuracy attainable with this device is proportional to the distance between the legs. Since a measurement cannot be made unless both legs rest on the surface being measured, however, we have several sizes of attachment, smaller ones for small surfaces, larger ones for larger surfaces.

To use the instrument we first read the dial gauge when the legs are upright on a flat surface. We then take a reading on the surface in question and write down the difference of the two readings. This difference is called s (for sagitta). We consider s as positive for convex surfaces, negative for concave surfaces.

To convert the reading of s into a radius of curvature, consider Fig. 12.2. Here balls of radius $b/2$ rest on a surface of radius R. The balls are a

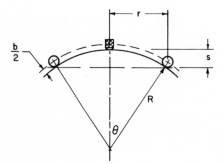

Figure 12.2 Deriving the spherometer equation.

distance $2r$ apart, and the dial gauge shows a difference s between this surface and a flat surface. Here

$$\left(R + \frac{b}{2}\right)^2 = r^2 + \left(R + \frac{b}{2} - s\right)^2$$

whence

$$2Rs = r^2 - bs + s^2$$

Thus

$$R = \frac{r^2}{2s} + \frac{s}{2} - \frac{b}{2}$$

Sometimes it is convenient to use this equation to evaluate r given R and *s.* The solution is

$$r = \sqrt{2Rs + bs - s^2}$$

Given R and r, we have

$$s = R + \frac{b}{2} - \sqrt{\left(R + \frac{b}{2}\right)^2 - r^2}$$

If $\sin \theta = r/(R + b/2)$, write

$$s = \left(R + \frac{b}{2}\right)(1 - \cos \theta)$$

If we keep track of the sign of s, these equations work equally well for concave and convex surfaces.

If the curvature of the surface is not uniform between the balls, the determination is of a mean curvature. A smaller spherometer could be used to measure curvature more locally.

The Reflection Method

If the surface to be measured is polished, its radius can be determined by observation of reflections. A simple system for use on highly curved objects appears in Fig. 12.3. Here a light source sends light through lens L. This light passes through the reticle R1 and is partially reflected by the half-reflecting mirror M down the microscope tube. The light strikes the microscope objective OB and is brought to a focus. If a reflecting spherical surface has its center of curvature at this point, the light is reflected back up as shown in details *a* and *b* for convex and concave surfaces, respectively. The objective OB sends the returning light through the half-reflecting mirror M and focuses it on the reticle R2. The fine focusing is done by means of the micrometer MS, the precision of focus being determined by the eye looking through the eyepiece EP. If the micrometer is now adjusted so that the reticle R1 is focused not at the center of curvature but on the spherical surface (detail *c*), rays passing through one side of lens OB return up through the opposite side and can again be focused on reticle R2, but the image is reversed. That is, if the opaque triangle of R1 appears below in one case, it appears above in the other. This opaque triangle is a help in finding the focus. Having found the triangle, we can concentrate on focusing the fine cross hair lines. The radius of curvature is the difference in micrometer readings between conditions *a* or *b* on the one hand and condition *c* on the other. In measuring convex surfaces the working distance of objective OB must exceed the radius of curvature of the surface, but there is no such

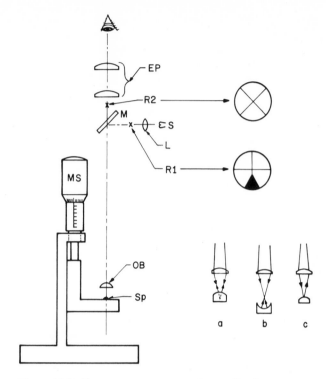

Figure 12.3   Curvature measurement for small spheres.

limitation in measuring concave surfaces. However the lens OB should be chosen with some care. For deep curves, greater precision follows a choice of a lens of high numerical aperture.

Surfaces that have different curvatures in different directions can be measured with this apparatus if reticle R1 has coarse "cross hairs." The doubly curved surface has a direction of maximum curvature and a direction of minimum curvature, and these directions are mutually perpendicular. If one of these directions is parallel to one of the cross hairs of R1, it is found to be possible to focus either hair of R1, but not both at the same time. This allows us to measure each radius of curvature separately in the same way as spherical surfaces were measured earlier.

The precision of this method is limited by the depth of focus of the microscope. Consider a wave front 1–3 (Fig. 12.4) coming from the point 0. If the point 0 were raised by amount Δ to coincide with point 2, the path difference at point 1 would be changed by amount ϵ. Applying the law of cosines to the triangle with sides R, Δ, and $R - \Delta + \epsilon$ gives

$$(R - \Delta + \epsilon)^2 = R^2 + \Delta^2 - 2R\Delta \cos \alpha$$

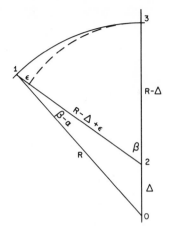

Figure 12.4   Analysis of depth of focus.

From this we get

$$\Delta = \epsilon\left(1 + \frac{\epsilon}{2R}\right)\Big/(1 - \cos\alpha) + \frac{\epsilon}{R} \simeq \frac{\epsilon}{1 - \cos\alpha}$$

where α is the angle of divergence of the light cone. Since phase to a quarter wave is the practical limit for resolution, we can take ϵ as $\lambda/4$ of the light used. Hence we find that a lens of numerical aperture N.A. has

$$\Delta = \frac{\lambda/4}{1 - \sqrt{1 - (\text{N.A.})^2}}$$

The focal depth is twice this, that is, Δ above the source and Δ below the source seem to be equally in focus.

The foregoing method gives a direct measurement of the radius of curvature. If the radius of curvature is rather large, the method is impractical. A method that depends on measuring the size of the image of some object is more appropriate in such a case.

Images in Spherical Mirrors

If we look at the image of a small spherical ball as seen reflected in a spherical mirror, we see not a circular image but an elliptical one. Consider a plane Q, defined by the three points: the observing eye, the ball center, and the center of curvature of the mirror. In a direction perpendicular to plane Q we find the long axis of the elliptical image of the ball, and in the plane Q we find the short axis of the ellipse. Hence we see that the magnification of the image is least in the plane Q, greatest perpendicular to the plane Q. We can determine the curvature of the mirror by measuring either of these two magnifications. We consider first magnification in the plane Q.

In Fig. 12.5 light from an object at O strikes the mirror at point 1 and is reflected along path (1–2). The angle of incidence is α. To an observer at point 2 the image of O appears to lie along the line (2–1). Light from O also strikes the mirror at point 3 and is reflected along the direction (3–4).

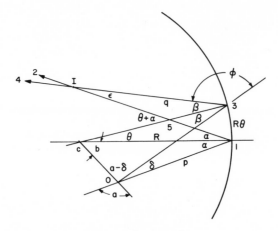

Figure 12.5 Images in a curved mirror—"in-plane" image.

The angle of incidence is β. To an eye at point 4 the image of point O appears to lie along the line (4–3). Hence the image of O is at I, the intersection of lines (2–1) and (4–3). Point c is the center of curvature of the mirror. The angle (103) is called δ, the angle (2I4) is called ϵ, and the angle (1c3) is called θ here. In triangle (01c), $a + b + \alpha = 180°$. In triangle (03c), $\alpha - \delta + b + \beta + \theta = 180°$. Hence $\beta = \alpha + \delta - \theta$. In triangle (513), $\alpha + \theta = \beta + \epsilon$. Hence $\epsilon = 2\theta - \delta$. Let $(01) = p$ and $(I1) = q$, the "object distance" and "image distance," respectively. In triangle (I13), by the law of sines, $\cos \beta = q \sin \epsilon/(R\theta)$, and in triangle (013), by the law of sines, $\cos \beta = p\delta/(R\theta)$. Since $\epsilon = 2\theta - \delta$ and $\theta = \delta/(R \cos \beta)$, we derive

$$\frac{1}{p} + \frac{1}{q} = \frac{2}{R \cos \beta}$$

The deviation ϕ of the light beam (due to reflection) is $180° - 2\beta$. Hence $\beta = 90° - \phi/2$, and we write

$$\frac{1}{p} + \frac{1}{q} = \frac{2}{R \sin (\phi/2)} \tag{12.1}$$

Since

$$\frac{1}{q} = \frac{2}{R \sin{(\phi/2)}} - \frac{1}{p}$$

we see that if p is less than $\frac{2}{R} \sin{\frac{\phi}{2}}$, then q must be negative; that is, the image is on the other side of the mirror, the side opposite the object side. This holds in general—if the image is on the side opposite the object side, q is negative. In the case of convex mirrors the image is always "inside," the object always "outside." Hence Eq. 12.1 holds for convex mirrors if q is taken as negative. This requires a convex mirror to have a negative radius and a negative focal length; in keeping with negative lenses, both show a diminished erect image.

We now treat the case of measurements perpendicular to the plane Q. Consider the narrow strip of curved mirror in Fig. 12.6, the axis of curvature being so labeled. Light from an object point at Ob strikes a small area of the mirror at the origin, with angle of incidence α. This ray

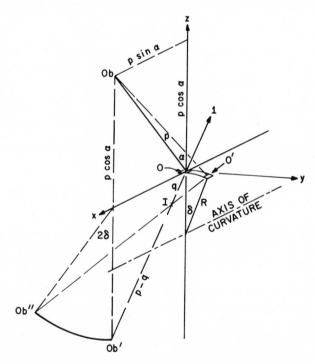

Figure 12.6   Images in a curved mirror—normal plane image.

reflects from the origin along the path $O-1$, O being the origin of the coordinate axes x, y, z. The ray seems to come from the direction of Ob′, the image of Ob as imaged in a flat mirror lying in the xy plane. Light from Ob also strikes an element of mirror centered at O'. This mirror element is tipped relative to the element at the origin, and it is tipped around the x axis by amount δ. Hence this bit of mirror images Ob at a point Ob″ which is Ob′ turned about the x axis by 2δ. Hence the image of Ob is at I intersection of lines (O, Ob′) and (O, Ob″) at a distance q from the origin.

The triangles (O, I, O') and (Ob', I, Ob'') are similar. Hence $(Ob', Ob''):(O, O') = (p - q):q$. But $(O, O') = R\delta$ and $(Ob', Ob'') = 2p\delta \cos \alpha$. Hence we have

$$\frac{1}{p} - \frac{1}{q} = \frac{-2 \cos \alpha}{R}$$

where q is negative, since it is inside the mirror, and R is negative, since the mirror is convex. Also the deviation angle $\phi = 180° - (\text{Ob}, O - O, 1)$; thus $\alpha = 90 - \phi/2$. Consequently we can write the equation for a concave mirror as

$$\frac{1}{p} + \frac{1}{q} = \frac{2 \sin \phi/2}{R} \qquad (12.2)$$

In summary of this: the object distance p, is always positive, whereas the image distance q is positive if the image lies on the same side of the mirror as the object; R is positive for concave mirrors.

We see that the two cases: "in the Q plane" and "perpendicular to the Q plane" differ only in the position of the term $\sin \phi/2$ (i.e., in the numerator or in the denominator).

MEASUREMENT OF IMAGES

We turn now to the actual determination of radii of curvature through measurement of images sizes. In the configuration of Fig. 12.7, both q and R are positive. We write the relevant equation as

$$\frac{1}{p} + \frac{1}{q} = \frac{2g}{R}$$

where g is $(\sin \phi/2)^{-1}$ for measurements in the plane of incidence (i.e., in the plane Q) but g is $\sin \phi/2$ for measurements perpendicular to the plane of incidence. The magnification is q/p and is generally less than unity. If q is negative the magnification is negative. Let the magnification for measurements in the plane of incidence be m and the magnification for measurements perpendicular to the plane of incidence be M. Both

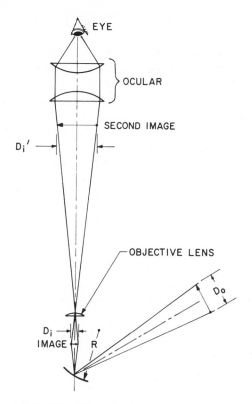

Figure 12.7   Measuring images.

measurements have the same value for p but have different values for q. For measurements in the plane of incidence we write

$$\frac{1}{p} + \frac{1}{mp} = \frac{2}{R \sin \phi/2}$$

hence

$$R = \frac{2mp}{(1+m)\sin(\phi/2)}$$

whereas for measurements perpendicular to the plane of incidence we have

$$R = \frac{2Mp \sin(\phi/2)}{1+M}$$

If the mirror is convex, the image is on the opposite side of the mirror, q

is negative, and so is the magnification. This results in a negative value for
R.

If the object is of dimension D_o and its image in the mirror is of
dimension D_i, the magnification is D_i/D_o with the proper sign assigned to
the value. With a traveling microscope it may be possible to measure D_i
directly. For very small images it is often preferable to measure the
second image by means of an eyepiece micrometer, which must be
"calibrated" for the ratio of D_i to D'_i, the dimension of the image in the
focal plane of the microscope objective.

Finally, in making such measurements of images to determine the
radius of curvature of a mirror, we can avoid the determination of ϕ by
combining the two sets of measurements—in the plane Q and the
perpendicular to the plane Q. This is done by taking the product of the
two equations for R. The result is

$$R^2 = \frac{4mMp^2}{(1+m)(1+M)}$$

The Eyepiece Displacement Method for Measuring Very Large Radii of Curvature

Consider a telescope (Fig. 12.8) consisting of an objective lens OB and an
eyepiece EP. We assume that the telescope is focused on a very distant
object and that the reticle R2 shows no parallax (i.e., the reticle markings
do not seem to move relative to the image of the distant object as the eye

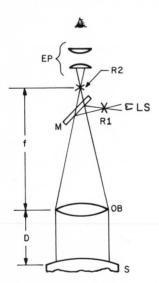

Figure 12.8 Measuring very large radii of curvature.

is moved across the eyepiece aperture). Hence the reticle accurately marks the position of the image of an object at "infinity." In this telescope there is another reticle R1, illuminated by a light source LS and a half-reflecting mirror M. With the telescope focused for infinity as just described, we place an optical flat below the objective so that light from the light source is reflected back through the telescope and an image of reticle R1 is formed near reticle R2. By adjusting the position of R1 we can make its image fall accurately in the plane of R2, as indicated by the lack of parallax between the two images. With R1 locked in this position we substitute a surface S of unknown spherical curvature at a distance D from OB and find that to focus R1 accurately we must move the eyepiece by amount Δ'. This motion is a measure of the curvature of surface S. To find the dependence of the radius of curvature of surface S on the value of Δ', we argue as follows. Light coming from R1 by way of mirror M is rendered parallel by objective OB. This light strikes surface S as though from infinity (i.e., the light wave front has zero curvature). By reflection from S the light will seem to come from a point $R/2$ below the surface, where R is the radius of curvature of surface S. The light reaches the objective as though from a distance $R/2 + D$ below OB; that is, the light has a wave front curvature $-(R/2 + D)^{-1} = -2/(R + 2D)$. The objective adds its "power" $1/f$ to change the wave front curvature to $1/f - 2/(R + 2D) = (R + 2D - 2f)/f(R + 2D)$, that is, this light comes to a focus at a distance $f(R + 2D)/(R + 2D - 2f)$ above OB. Hence $\Delta' = 2f^2/(R + 2D - 2f)$.

If now we make $D = f$, we have the simple expression

$$\Delta' = \frac{2f^2}{R}$$

that is,

$$R = 2f^2/\Delta'.$$

For example, if $f = 20$ cm and $\Delta' = 0.1$ cm, we find $R = 80$ meters.

Although the method of Fig. 12.3 can be applied for large radii, using an optical bench instead of a micrometer screw to measure the distance between the two focusing positions, an optical bench capable of measuring 80 meters is not very practical. Hence the importance of the eyepiece displacement method.

Angle Measurement

The autocollimator can be used to measure the angle between the two sides of an almost parallel sided transparent slab. Figure 12.9 supplies the details. Here light strikes the upper surface of a transparent slab perpendicularly and is partially reflected back along its incoming path.

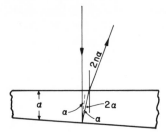

Figure 12.9 Measuring angles between plate surfaces.

The transmitted portion strikes the lower surface of the slab at an angle of incidence α, where α is the angle between the surfaces. The light reflected from the lower surface arrives at the upper surface with an angle of incidence 2α and is refracted at an angle $2\alpha n$ from the perpendicular, where n is the refractive index of the slab. Hence in the eyepiece we see two images of the reference reticle. The separation of these in the focal plane of the autocollimator objective is $2\alpha nf$, where f is the focal length of the objective lens of the autocollimator. The α should be measured in radians, but the sine or tangent of α measured in any units is nearly the same thing. As an example let us assume $f = 20$ cm, $n = 1.5$, $\alpha = 1$ minute. Hence the images are separated by 0.01746 cm. If the eyepiece magnifies 10 times, this separation looks like 1.7 mm. We must consider the resolution of the system if we attempt to determine very small angles. A telescope with an objective of diameter D can resolve two points as two distinct points if their angular separation exceeds $1.22\lambda/D$ radians, where λ is the wavelength of the light. If the slab is of limited size (i.e., smaller than the objective lens), we should use its dimension in place of D. In the example a one-minute resolution would require the "slab" to have a diameter of at least 0.2 mm. A resolution of one second requires a 12-mm slab, both of these on the assumption that $\lambda = 0.5\ \mu$ and that the objective is perfect.

In measuring angles between polished flat opaque surfaces, a spectrometer serves well as a goniometer. However an autocollimator and a set of standard angle blocks are better. Blocks accurate to better than a second are commercially available. These can be wrung together as is done with gauge blocks. If all the big ends are kept on one side, the result is the sum of the individual angles; but blocks can be reversed, whereupon they subtract from the total of the unreversed blocks. In this way 13 blocks can be used to make any angle up to 90° within a second of arc. This minimal set is 1″, 3″, 9″, 27″, ... However it is not easy to see just what blocks are needed to make some arbitrary angle. Also angles that are submultiples of 360° are most commonly needed. Hence a practical set is the 16 blocks 1″, 3″, 5″, 20″, 30″, 1′, 3′, 5′, 20′, 30′, 1°, 3°, 5°, 15°, 30°, 45°.

With the minimal set there is only one way to construct an angle, but with the 16-block set there are several ways for many required angles. Because each block has a probable error and these probable errors are roughly the same for all, an angle is best constructed with the least possible number of pieces.

Given an arbitrary angle, we can treat the degrees, minutes, and seconds separately, finding combinations of pieces that will produce the degrees, the minutes, and the seconds, then combining these to produce the required angle to one second. Table 12.1 lists combinations that can be used to make the seconds using second blocks or to make the minutes

TABLE 12.1   SECOND AND MINUTE COM-
BINATIONS

1	$31 = 30 + 1$
$2 = 3 - 1$ or $5 - 3$	$32 = 30 + 5 - 3$
$3 = 3$	$33 = 30 + 3$
$4 = 5 - 1$	$34 = 30 + 5 - 1$
$5 = 5$	$35 = 30 + 5$
$6 = 5 + 1$	$36 = 30 + 5 + 1$
$7 = 5 + 3 - 1$	$37 = 30 + 5 + 3 - 1$
$8 = 5 + 3$	$38 = 30 + 5 + 3$
$9 = 5 + 3 + 1$	$39 = 30 + 5 + 3 + 1$
$10 = 30 - 20$	$40 = 60 - 20$
$11 = 30 - 20 + 1$	$41 = 60 - 20 + 1$
$12 = 20 - 5 - 3$	$42 = 60 - 20 + 5 - 3$
$13 = 30 - 20 + 3$	$43 = 60 - 20 + 3$
$14 = 30 - 20 + 5 - 1$	$44 = 60 - 20 + 5 - 1$
$15 = 30 - 20 + 5$	$45 = 60 - 20 + 5$
$16 = 20 - 3 - 1$	$46 = 60 - 20 + 5 + 1$
$17 = 20 - 3$	$47 = 30 + 20 - 3$
$18 = 20 + 1 - 3$	$48 = 30 + 20 + 1 - 3$
$19 = 20 - 1$	$49 = 30 + 20 - 1$
$20 = 20$	$50 = 30 + 20$
$21 = 20 + 1$	$51 = 30 + 20 + 1$
$22 = 20 + 3 - 1$	$52 = 30 + 20 + 5 - 3$
$23 = 20 + 3$	$53 = 30 + 20 + 3$
$24 = 20 + 5 - 1$	$54 = 30 + 20 + 5 - 1$
$25 = 20 + 5$	$55 = 30 + 20 + 5$
$26 = 20 + 5 + 1$	$56 = 30 + 20 + 5 + 1$
$27 = 30 - 3$	$57 = 60 - 3$
$28 = 30 - 3 + 1$	$58 = 60 - 3 + 1$
$29 = 30 - 1$	$59 = 60 - 1$
$30 = 30$	

using the minute blocks. Table 12.2 does the same for degrees. The 210 entries required to accomplish this would require 324,000 entries with the minimal set of 13 blocks. The table would be even shorter if we had a 90° block and entered degrees only, to 45°. We could then build back from 90° in such an obvious way that a table beyond 45° would not be required. In the autocollimator setup in Fig. 12.10, a micrometer eyepiece is used to measure deviations from the standard.

There is a classically simple way of testing 90° by means of an autocollimator. Figure 12.11 presents the scheme. Light from the autocollimator reticle strikes the surface of the plate P at an angle and is reflected onto the block B, from which it reflects back upward. The block angle is here marked as $90° - \alpha$, and if α is zero, the returning ray is parallel to the downward ray; otherwise it makes an angle 2α with the downward ray and an image appears in the eyepiece field displaced by amount $2\alpha f$ to the right of where it would be if $\alpha = 0$. Moreover, a downward ray (dashed in Fig. 12.10) that strikes the block first reflects onto the plate and then goes upward, making an angle 2α to the left. Hence we see two images, and

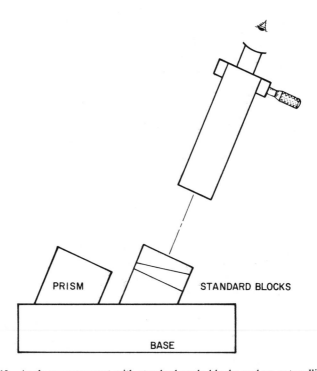

Figure 12.10 Angle measurement with standard angle blocks and an autocollimator.

TABLE 12.2   DEGREE COMBINATIONS

1	$46 = 45 + 1$
$2 = 3 - 1$	$47 = 45 + 3 - 1$
3	$48 = 45 + 3$
$4 = 3 + 1$	$49 = 45 + 3 + 1$
5	$50 = 45 + 5$
$6 = 5 + 1$	$51 = 45 + 5 + 1$
$7 = 5 + 3 - 1$	$52 = 45 + 5 + 3 - 1$
$8 = 5 + 3$	$53 = 45 + 5 + 3$
$9 = 5 + 3 + 1$	$54 = 45 + 5 + 3 + 1$
$10 = 15 - 5$	$55 = 45 + 15 - 5$
$11 = 15 - 5 + 1$	$56 = 45 + 15 - 5 + 1$
$12 = 15 - 3$	$57 = 45 + 15 - 3$
$13 = 15 + 1 - 3$	$58 = 45 + 15 - 3 + 1$
$14 = 15 - 1$	$59 = 45 + 15 - 1$
15	$60 = 45 + 15$
$16 = 15 + 1$	$61 = 45 + 15 + 1$
$17 = 15 + 3 - 1$	$62 = 45 + 15 + 3 - 1$
$18 = 15 + 3$	$63 = 45 + 15 + 3$
$19 = 15 + 3 + 1$	$64 = 45 + 15 + 5 - 1$
$20 = 15 + 5$	$65 = 45 + 15 + 5$
$21 = 15 + 5 + 1$	$66 = 45 + 15 + 5 + 1$
$22 = 15 + 5 + 3 - 1$	$67 = 45 + 30 - 5 - 3$
$23 = 15 + 5 + 3$	$68 = 45 + 15 + 5 + 3$
$24 = 15 + 5 + 3 + 1$	$69 = 45 + 30 - 5 - 1$
$25 = 30 - 5$	$70 = 45 + 30 - 5$
$26 = 30 - 5 + 1$	$71 = 45 + 30 - 5 + 1$
$27 = 30 - 3$	$72 = 45 + 30 - 3$
$28 = 30 + 1 - 3$	$73 = 45 + 30 - 3 + 1$
$29 = 30 - 1$	$74 = 45 + 30 - 1$
30	$75 = 45 + 30$
$31 = 30 + 1$	$76 = 45 + 30 + 1$
$32 = 30 + 3 - 1$	$77 = 45 + 30 + 3 - 1$
$33 = 30 + 3$	$78 = 45 + 30 + 3$
$34 = 30 + 5 - 1$	$79 = 45 + 30 + 5 - 1$
$35 = 30 + 5$	$80 = 45 + 30 + 5$
$36 = 30 + 5 + 1$	$81 = 45 + 30 + 5 + 1$
$37 = 45 - 5 - 3$	$82 = 45 + 30 + 5 + 3 - 1$
$38 = 45 - 5 + 1 - 3$	$83 = 45 + 30 + 5 + 3$
$39 = 45 - 5 - 1$	$84 = 45 + 30 + 5 + 3 + 1$
$40 = 45 - 5$	$85 = 45 + 30 + 15 - 5$
$41 = 45 - 5 + 1$	$86 = 45 + 30 + 15 - 3 - 1$
$42 = 45 - 3$	$87 = 45 + 30 + 15 - 3$
$43 = 45 - 3 + 1$	$88 = 45 + 30 + 15 - 3 + 1$
$44 = 45 - 1$	$89 = 45 + 30 + 15 - 1$
45	$90 = 45 + 30 + 15$

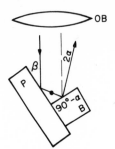

Figure 12.11   Checking a 90° angle.

they are separated by $4\alpha f$ in the focal of the objective OB. If $f = 20$ cm and $\alpha =$ one minute of arc, the images are separated by 0.23 mm and a ten-power eyepiece makes this look to the eye like 2.3 mm. Hence we can make 90° blocks within seconds of arc even with an autocollimator of only 20 cm focal length. It should be mentioned that 45° can be checked in the same way. The 45° angle has four reflections, and the images are separated by $8\alpha f$. However the images become faint with so many reflections.

A device for checking squareness of small unpolished plates appears in Fig. 12.12. A very square block C is mounted on two pivots. The device rests on an autocollimator base plate, and a crystal plate is slid along bar B until it presses against block C. This forces the block so that the upper face is parallel to the autocollimator base if the crystal angle is 90°. If the angle is not 90°, the error can be read immediately.

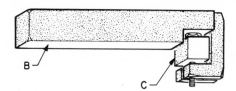

Figure 12.12   Tool for checking squareness of unpolished plates.

The Anamorphotic Microscope

In observing small deviations from straightness, a microscope that magnifies differently in different directions can be employed. Figure 12.13 reveals that the "objective" is made up of two cylindrical lenses 01 and 02. The cylinder axes of the lenses are set at 90° each to each. Element 01 magnifies distances along the x direction, projecting an image on plane I. Element 02 projects a reduced image of the object plane onto the image plane I. Care must be taken to make these two images coincide. When the

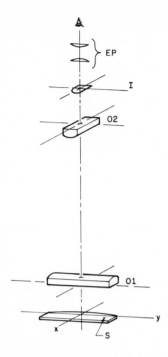

Figure 12.13   The anamorphotic microscope.

instrument is properly adjusted, an exaggerated image is produced: the slight curve of an edge of sample S is projected as a very deep curve, as illustrated in the image plane of Fig. 12.13. The instrument can cover a large field along y—we have covered 2 in. in one view, the element 01 being 2 in. long. An ordinary eyepiece observes the image, and a standard micrometer eyepiece permits measurements to be made. If the elements are not color corrected, monochromatic light should be used.

The Profile Microscope

In making small tools such as laps for very small concave spherical surfaces, one may wish to inspect them under a microscope. Figure 12.14 shows a microscope used for this purpose. A microscope objective MO casts a magnified image of the point of the tool T in the focal plane FP. A slot through the microscope barrel allows the insertion of a transparent slide in this focal plane. Such slides are generally made of plastic and have reference figures inscribed on them. Circles can be inscribed readily by the use of a draughtsman's drop-spring bow compass, simply replacing the lead with a steel scriber. For more complicated designs, a large drawing photographically reduced to the proper size serves nicely. For

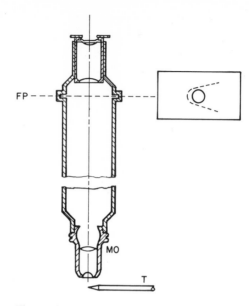

Figure 12.14 The profile microscope.

example, a tool required to lap a concave spherical surface to a radius of 0.1 mm should be honed until the end matches a 2-mm-radius circle if a 20X microscope objective is used. Since the tube length affects the actual magnification, it is wise to measure the image of a small wire. This can be done with a steel scale in the slot to sufficient accuracy. The slide should be enough narrower than the slot to permit the slide to be slid both lengthwise and crosswise to match the image. With a 20X objective it is 20 times easier to position the slide to a match than it is to position the tool.

A Laser Instrument to Determine Errors in Parallelism of Rounded Ends

The device in Fig. 12.15 is for testing parallelism of rounded ends. A helium-neon laser beam reflects from a rounded end and falls on a screen as a streak, which can be centered between vertical lines drawn on the screen. A micrometer attached to the fixed bar FB bends the flat spring FS, thus swinging the movable bar MB. The crystal is pressed against MB as the micrometer is adjusted to place the reflection symmetrically between the lines ruled on the screen. The fixed bar FB has an effective length of $3\frac{7}{16}$ in. which makes one division of the micrometer (0.001 in.) equivalent to a minute of arc. A standard right angle piece STD is used to find the reading for 90°. One face is optically flat with a mirror finish. The

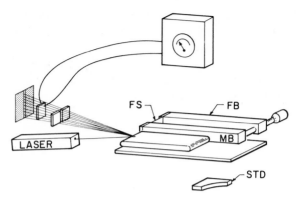

Figure 12.15 Laser for checking angles between rounded ends.

right angle is checked by the method of Fig. 12.11. Greater accuracy of setting is obtained by detecting the reflected light by means of a photodetector (a silicon cell does very nicely). The cell is in a holder in a metal block, the light entering through a slit just wide enough to pass the beam. This arrangement reduces the troublesome effect of room light.

The Divergent Beam Interferometer

The laser is useful for several tasks because of the large path differences over which interference may be observed. Figure 12.16 is a schematic view of a divergent beam interferometer. The laser beam, which is highly parallel, is rendered divergent by lens L_1 and is reflected down by the beam splitter BS. After reflection at the two surfaces of the object being

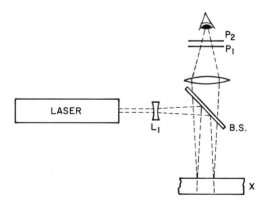

Figure 12.16 Divergent beam interferometer.

tested, the interfering beams pass through the beam splitter and are converged to the observing eye. The polarizers P_1 and P_2 dim the light, preventing injury to the eye.

Without the beam splitter, polarized laser light can be used with the lenses in a line with the laser beam to make a conoscope that will show fringes for thick crystals far off the optic axis.

13

Miscellaneous

PRECISION X-RAY GONIOMETRY

To achieve orientations more accurate than 3 or 4 minutes of arc with a single crystal goniometer requires special care in instrument design, as well as fairly perfect crystals. A strained crystal gives broad reflections. Cutting or grinding the crystal surface causes strains. These defects may be relieved by etching. Ductile crystals may be deeply strained and may require the removal of much material. Even quartz, a far from ductile material, may have its double crystal rocking curves narrowed by a factor of 10 by a brief etch in hydrofluoric acid. Given unstrained crystals, we must still consider the angular spread of rays coming out of the collimator and also the spread of wavelengths that constitute the $K\alpha$ radiation. We assume that a nickel filter has removed the $K\beta$ radiation.

Wavelength Spread

The wavelength spread is a double problem because there are two wavelengths quite close together; that is, $K\alpha$, is a "doublet." For copper radiation $K\alpha_1$ has a wavelength of 1.540562 Å, whereas $K\alpha_2$ has a wavelength of 1.544398 Å, a difference of 0.003836 Å. Moreover neither line is perfectly sharp. As Fig. 13.1 indicates, the $K\alpha_1$ line has a width of 0.00058 Å at half-maximum and the $K\alpha_2$ line has a width of 0.00077 Å at half-maximum, since $\lambda = 2d \sin \theta$ differentiation shows that variations in λ cause variations in θ according to the equation $\Delta\theta = (\Delta\lambda/\lambda) \tan \theta$ radians $= 3438(\Delta\lambda/\lambda) \tan \theta$ minutes. Hence a crystal plane for which the Bragg angle is 20° for copper $K\alpha_1$ will have a Bragg angle of $20° + 3438 \times 0.003836 \tan 20°/1.54$ minutes for $K\alpha_2$, that is, 3.12 minutes between peaks. Moreover the width at half-maximum for $K\alpha_1$ is $3438 \times 0.00058 \tan 20°/1.54 = 0.47$ minutes, whereas for $K\alpha_2$ this width is 0.62 minutes. If the collimator spread is less than 3 minutes, $K\alpha_1$ and $K\alpha_2$ would be resolved at 20°. The response for $K\alpha_1$ will be about twice as strong as that for $K\alpha_2$ and it will appear at the smaller Bragg angle.

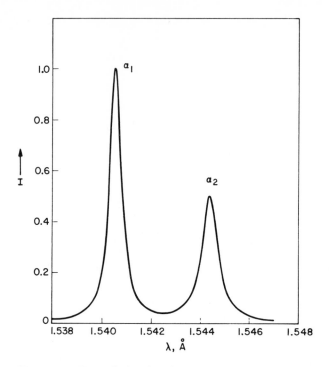

Figure 13.1   Kα radiation lines for copper.

Most X-ray lines have greater values of $\Delta\lambda/\lambda$ than does copper. However there is a slight gain with zinc, which is 94% as wide as copper, and germanium, which is 91% as wide as copper. This is not gain enough to pay for the trouble of changing X-ray tubes. The L series of the heavy elements have wavelengths in the useful range but have broader lines. Tungsten Lβ_1 with a wavelength 1.281809 Å has a $\Delta\lambda/\lambda$ value about twice that of copper. There is an advantage in using molybdenum for very small unit cells. For Mo we have λ Kα_1 = 0.707831 Å, $\Delta\lambda/\lambda$ = 0.000205. For very large unit cells iron presents an advantage, since λ Kα_1 = 1.932026 Å, $\Delta\lambda/\lambda$ = 0.000259.

Collimator Spread

A collimator with slits of width w separated by a distance l gives a beam spread of $\tan^{-1}(2w/l)$, that is, $6876w/l$ minutes, neglecting a secondary effect due to vertical spread. If the backward-projected limiting rays of the collimator all fall on the radiating area of the X-ray tube target, the effective reflected radiation spread is the sum of the linewidth spread and the collimator spread. Hence to resolve the doublet we need a collimator spread less than the doublet spread. If the doublet is not resolved, the effective wavelength is not that of $K\alpha_1$ or $K\alpha_2$ but some unknown intermediate value. Bragg angles cannot be calculated as accurately with such an unknown compromise wavelength as with the more sharply defined wavelengths, preferably that of $K\alpha_1$. Crystals with large cells have many Bragg angles, and proper identification of these requires careful measurement.

To separate the copper $K\alpha$ doublet at 20° requires a collimator equivalent to two slits of width 0.05 mm separated by 110 mm. Finer slits with greater separation can be used, but the counting tube receives weaker, hence more erratic signals; it is "jumpy." Erratic readings are due to the statistical nature of the quanta of X-radiation; in small samples variation is more obvious than in large samples.

Intensity Measurements

For crystal orientation work a Geiger tube has the advantages of sensitivity, as compared to an ionization chamber, and simplicity of electronic equipment, as compared to scintillation counters or proportional counters. Because of loss of recovery time of avalanche breakdown, however, the Geiger tube is not linear in its response. Geiger tubes react differently. A rough calibration is afforded by observing the X-ray tube accelerating potential as a function of the apparent counting rate as read from the counting rate meter. Between 8 and about 40 kV the output of $K\alpha$ radiation is approximately proportional to

$$\left(\frac{kV}{8} - 1\right)^{1.65}$$

There is no $K\alpha$ radiation below 8 kV. (The reported actual limiting value is 8.04778 kV but 8 kV is sufficient here.)

People are sometimes puzzled because a crystal reflects more strongly if it is turned 180° about the reflecting plane normal. The crystal surface is not parallel to the atomic reflecting planes, however, and rays that impinge on the crystal surface at a small grazing angle but leave at a larger grazing angle are reflected more copiously than rays impinging at a larger grazing angle but departing at a smaller grazing angle. Rays arriving at smaller angles cover a larger area of crystal surface but reflect from shallower depths than rays arriving at larger grazing angles. A crystal that reflects with an intensity I_0 with a Bragg angle θ when the surface is parallel to this atomic reflecting plane will exhibit a different intensity I from the same atomic plane when the surface is not parallel to this atomic plane. Here if Δ is the angle between plane and surface, we have

$$I \simeq \frac{I_0 \sin \theta}{\sin (\theta + \Delta)}$$

If Δ is positive, I is less than I_0, if Δ is negative, I is larger than I_0. For example, with $\theta = 20°$ and $\Delta = 10°$, $I_+ = I_0 \sin 20°/\sin 30° = 0.681 I_0$, whereas $I_- = I_0 \sin 20°/\sin 10° = 1.97 I_0$. Hence on turning this crystal around we should expect the intensity to change by the ratio $\sin 30°/\sin 10° = 2.88$. With very small grazing angles, and if the counting tube window is small, not all the reflected rays may be able to enter the counting tube, giving an erroneous answer.

The Double Crystal Goniometer for Precision Orientation

For very precise orientation of crystals it is possible to use a double crystal goniometer. The first crystal should have a lattice spacing matching that of the second crystal (i.e., the one being oriented). This limits the versatility of the instrument. If one has on hand a half-dozen "perfect" crystals with different Bragg angles to be used as the first crystal (crystal A of Fig. 4.34), one can do a fair job of covering any desired Bragg angle. However each A crystal must be very carefully adjusted before use. This makes the double crystal instrument good for repetitive work such as manufacturing one standard orientation from one kind of crystal. It is impractical for general work.

As mentioned in Chapter 6, the reflected radiation angular spread is inversely proportional to $\sin 2\theta$, and with "perfect" crystals it may be only a few seconds wide at $\theta = 45°$, the sharpest position. If the width at $\theta = 45°$ is 10 seconds, it will be 20 seconds wide at $\theta = 15°$ and also at $\theta = 75°$.

One difference between double and single crystal goniometers is in the area of crystal covered by radiation. The single crystal goniometer orients a crystal by "sampling" a small part of the crystal area. The double crystal goniometer "samples" a much larger part of the area. This could conceivably be an advantage in some cases, a disadvantage in others.

Lapping Very Precisely Oriented Surfaces

For the greatest precision of orientation, the surface must be very flat. A 1-in. diameter surface that is 0.001 in. high in the center may be considered as part of a sphere of radius 125 in. Opposite edges differ in orientation by more than a degree of arc. To lap surfaces flatter, opticians have long cemented the object to the center of a substantial block and surrounded it with pads of about the same hardness and thickness (see Fig. 13.2). The assembly is ground and lapped on very flat laps and

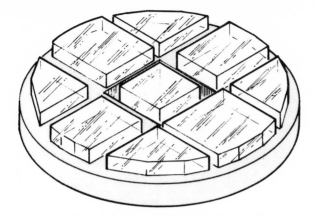

Figure 13.2 Object to be lapped is surrounded with pads of similar hardness and thickness.

polished on a very flat polishing lap. We assume that the initial orientation is within, say, a half degree. The assembly is X-ray checked by pressing the pads against a flat reference surface that has a central hole through which X-rays pass coming and going as shown in Fig. 13.3. The assembly rests in vees and is pressed gently against the reference surface. Turning the assembly 180° about its cylinder axes allows one to measure the surface deviation from an atomic reference plane. Four observations 90° apart give a complete check. Orientation correction can be made by bringing extra pressure to bear off center while grinding or polishing. The off-center pressure is applied in such a way as to correct the orientation.

For the ends of longer bars the composite mounting block of Fig. 13.4 is convenient. Mounted on a dovetail slide, the crystal can be adjusted up

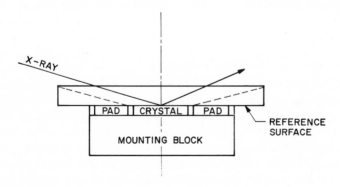

Figure 13.3 X-ray checking of assembly in Fig. 13.2.

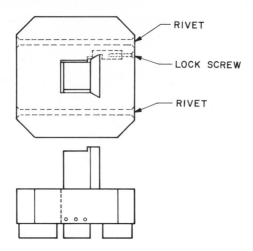

Figure 13.4   Composite mounting block for ends of longer bars.

and down. The slide can be removed from the assembly, making it easy to cement the crystal to it.

A third precision fixture is illustrated in Fig. 13.5. The crystal is mounted on a dovetail slide but the slide is attached to a floating block whose spherical bottom fits into a spherical seat of the lower base block to which the pads are cemented. Four screws (only two shown) move the floating block about on its seat, making slight angular adjustments. Since the seat radius is rather large, the adjustment is quite sensitive.

In another model (Fig. 13.6) the float block is in two pieces. One piece is forced back and forth along a single line by two screws but supports two screws that force an inner block along the curve. A gimbal device for the same purpose appears in Fig. 13.7. The reference surface of Fig. 13.2 is at the left.

Precise Orientation by Laue Photographs

With the devices of Figs. 13.4 to 13.7, it is possible to use an accurately made Laue camera to achieve orientations to better than a minute of arc if there are several symmetry-related reflections on the photograph. The Laue camera must have a flat reference surface against which the pads (Fig. 13.5) may rest. The X-ray beam from the camera collimator must be perpendicular to this reference plane to a fraction of a minute of angle. In use, the crystal is adjusted until symmetry-related spots show the same intensity on the photograph.

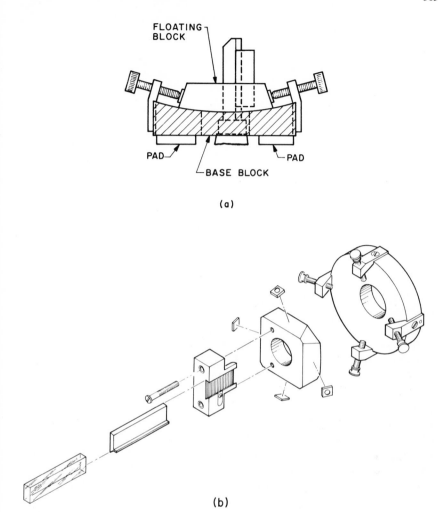

FLOATING BLOCK

PAD

PAD

BASE BLOCK

(a)

(b)

Figure 13.5 Precision fixture, slide attached to floating block. (*a*) Cross section; (*b*) exploded view.

Refraction

A source of slight error in angle measurement is the refraction bending of the X-rays as they enter or leave the crystal. Bragg's law holds inside the crystal, but X-ray wavelengths are given for "free space." The wavelengths are slightly longer inside the crystal—namely, $\lambda = \lambda_0/n$, where n is the refractive index and λ_0 is the wavelength in free space. The index of refraction is slightly less than 1. We write $n = 1 - \delta'$, where δ' is

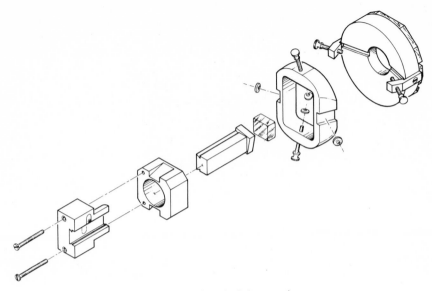

Figure 13.6 Precision fixture with floating block in two pieces.

of the order of 10^{-5}. If we sum the atomic numbers of all atoms in a unit cell and divide this by the volume of the unit cell in cubic angstroms we get a number n_0, the number of orbital electrons in a cubic angstrom. It can be shown that $\delta' \sim 4.48 \times 10^{-6} n_0 \lambda_0^2$, where λ_0 is expressed in angstroms.

The refraction error is greatest for small grazing angles. The grazing angle g is related to the angle of incidence i, used in ordinary optics as $g = 90° - i$. Hence the law of fraction may be written as $\cos g_i = n \cos g_x$, where g_x is the grazing angle inside the crystal and g_i, the grazing angle outside, is slightly larger (by the amount ε). Hence $\cos(g_x + \varepsilon) = (1 - \delta) \cos g_x$. Expanding and reducing this we derive $\varepsilon = \delta / \tan g_x$.

We illustrate this with the case of bismuth germanium oxide (BGO). There are two molecules in a cubic unit cell 10.1455 Å on a side. Hence there are 24 bismuth atoms, each of which has 83 electrons, 2 germanium atoms with 32 electrons each, and 40 oxygen atoms with 8 electrons each, a total of 2376 electrons in a 10.1455 Å cube (i.e., 2.28 electrons per cubic angstrom). Hence here $\delta = 24 \times 10^{-6}$. For copper K$\alpha$ radiation, the 110 plane of BGO has a Bragg angle of 6°10'. If these atomic planes are parallel to the crystal surface, the Bragg angle will appear larger than 6°10' by the amount $\varepsilon = (24 \times 10^{-6} / \tan 6°10')$ radian, that is, by $(3438 \times 24 \times 10^{-6} / \tan 6°10')$ minute, or 0.76 minute. If the atomic planes are not parallel to the crystal surface but are tipped by, say, 4 degrees, giving an incident grazing angle of 2°10', then ε will be 2.2 minutes, which is a noticeable error.

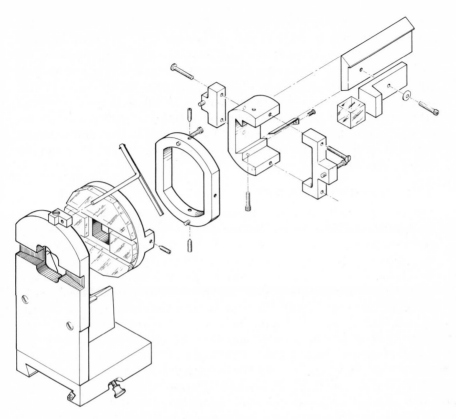

Figure 13.7   Gimbal device for ensuring precision.

How Many Reflections Are There?

By counting the nodes in the limiting sphere such as that of Fig. 4.6 we find that for a primitive cell, in which all values of h, k, and l are valid, there are $[32\pi/(3\lambda^3)] \times$ volume of unit cell reflections possible. For nonprimitive cells there are far fewer possible reflections. For example, molybdenum, a body-centered cubic structure with $a_0 = 3.140$ Å, would have about 284 reflections if it were primitive; since it is body centered, however, it has only about half this number. Many of these planes are merely a shuffling of indices such as $hkl, khl, \ldots, \bar{h}kl \ldots \bar{h}\bar{k}\bar{l}$. Hence the number of distinct Bragg angles is rather small.

X-Ray Density

Given the chemical formula for a crystal and the cell constants, including the number of molecules in the unit cell, we can calculate the mass density. The equation is $\rho = (1.6603 \, \Sigma M)/V$, where ΣM is the sum of the molecular weights in the unit cell of volume $V \, \text{Å}^3$. We illustrate with germanium. There are eight atoms of atomic weight 72.6 in a cubic cell 5.6575 Å on a side. Hence the density should be 5.325 g/cm^3. However older tables give the density as 5.35 g/cm^3. The older data must have come from measurements on impure germanium, the impurity being a heavier element. However even pure substances may not exhibit exactly the density as just calculated because of vacancies. The foregoing equation can be used to calculate the number of atoms in a unit cell, given the density and the size of the unit cell.

X-Ray Absorption Calculations

The loss of X-ray intensity caused by passage through an absorbing medium is calculated by use of the equation $I = I_0/\exp(\mu t)$, where μ is the linear absorption coefficient and t is the path length. Tables giving for the elements the *mass* absorption coefficient μ/ρ for many X-ray wavelengths can be found in many books on X-rays. These values, if multiplied by the density ρ, give the linear absorption coefficient μ.

For example, at $\lambda = 1.54$ Å aluminum has a mass absorption coefficient of 49.0. With a density 2.702 the linear absorption coefficient is 132.4/cm. Hence a sheet of aluminum 0.1 mm thick reduces X-rays of wavelength 1.54 Å in the ratio $\exp(1.324) = 3.76$; that is, 27% passes through. However a 1-mm sheet passes 0.000178%. Copper, with $\mu/\rho = 50.9$, $\rho = 8.92$, has a linear absorption coefficient of 454/cm, and 0.1 mm of copper passes 1.07% of copper Kα radiation. We choose brass to illustrate the method of calculating absorption of nonelementary materials. Common brass is 67% copper by weight and 33% zinc; its density is 8.40 g/cm^3. The respective densities of the copper and the zinc in unit volume of this are $0.67 \times 8.40 = 5.628$ g/cm^3 and $0.33 \times 8.40 = 2.772$ g/cm^3. The linear absorption coefficient of this brass is then $5.628(\mu/\rho)_{Cu} + 2.772(\mu/\rho)_{Zn} = 5.628 \times 50.9 + 2.772 \times 58.6 = 448.9$, a trifle less than that for pure copper. A tenth of a millimeter of this brass passes 1.12% of copper Kα radiation.

Counting Rate Meter Calibration by Means of Foils

With a number of foils of known composition and thickness, we can calibrate a counting rate meter. For example, with 0.002-in. aluminum foils the following table would apply

Number of foils	1	2	3	4	5	6	7	n
% transmission	51.0	26.0	13.3	6.78	3.46	1.77	0.902	$100(0.51^n)$

With copper radiation we do not use iron or steel foils because iron fluoresces under this X-ray wavelength. In fact in doing orientation work it is well to prevent the beam from falling on iron because the fluorescent radiation, which is also a health hazard, loads the radiation detector.

Absorption Edges

The absorption of X-rays tends to be greater for longer wavelengths and, if plotted for a limited range, seems to obey a cube law—that is, μ/ρ is proportional to λ^3. If absorption is plotted for a longer range, however, discontinuities appear just above the $K\beta$ wavelength of the absorber. As λ is increased past the $K\beta$ wavelength, the mass absorption coefficient falls precipitously, then rises again following the cube law. In the curve for nickel (Fig. 13.8) $\log \lambda$ is plotted against $\log \mu/\rho$. It is obvious here why a nickel filter substantially reduces the copper $K\beta$ radiation without greatly reducing the copper $K\alpha$ radiation.

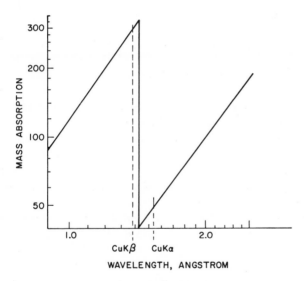

Figure 13.8 Mass absorption curve for nickel, showing the discontinuity between the wavelength for copper $K\beta$ and copper $K\alpha$.

CALCULATING SOME PHYSICAL PROPERTIES

In crystal calculations a working knowledge of vectors and matrices simplifies the solution of problems. Let us now present the fundamental ideas and explain the operations in sufficient detail to enable the reader to handle such problems. A scalar is a mere number, such as 7, π, or a single algebraic variable or constant.

Vectors

A vector is a directed quantity; it has length and direction. It is always stated on a coordinate system such as the x, y, z axes of an orthogonal system. We write the coordinates in a single column as

$$\begin{pmatrix} x \\ y \\ z \end{pmatrix}$$

To save space we often write this as a row $[xyz]$, enclosed in square brackets to indicate that it is a vector written in a row. In multiplying a scalar quantity and a vector, each element of the vector is multiplied by the scalar:

$$u \begin{pmatrix} x \\ y \\ z \end{pmatrix} = \begin{pmatrix} ux \\ uy \\ uz \end{pmatrix} \qquad \text{is } u \text{ times as long as} \qquad \begin{pmatrix} x \\ y \\ z \end{pmatrix}$$

but has the same direction in space as $[xyz]$. In rectangular coordinates the length of vector $[xyz]$ is $\sqrt{x^2 + y^2 + z^2}$. If we multiply the vector by the reciprocal of its length, we make a new vector along the old vector but of unit length.

The scalar product of two vectors, which is the product of their lengths times the cosine of the angle between them, is not a vector but a mere number. Operationally the scalar product of two vectors $[xyz]$ and $[x'y'z']$ is the quantity $xx' + yy' + zz'$. A dot between two vectors indicates the scalar product of them; consequently two unit vectors, when multiplied scalarly, give the cosine of the angle between them.

The *vector product* of two vectors is defined as a third vector perpendicular to both. Its length is the product of the lengths of the two vectors times the sine of the angle between them. Its direction is given by the right-hand rule: $a \times b = c$: if the right-hand index finger points from the origin toward the end of a and the middle finger points from the origin in the direction of b, the thumb points in the direction of c. Consequently $a \times b = -b \times a$. The cross signifies the vector product.

Operationally we have

$$\begin{pmatrix} a_1 \\ a_2 \\ a_3 \end{pmatrix} \times \begin{pmatrix} b_1 \\ b_2 \\ b_3 \end{pmatrix} = \begin{pmatrix} a_2 b_3 - a_3 b_2 \\ a_3 b_1 - a_1 b_3 \\ a_1 b_2 - a_2 b_1 \end{pmatrix}$$

Elementary crystallography texts form this product by writing each vector twice, one under the other, then striking out the two end members and cross multiplying the rest. As written out, the result is neater if we write the two vectors each twice in a single column, then strike out the top and bottom and cross multiply the rest

$$
\begin{array}{cc}
a_1 & b_1 \\
a_2 & b_2 \\
a_3 & b_3 \\
a_1 & b_1 \\
a_2 & b_2 \\
a_3 & b_3
\end{array}
\quad
\begin{aligned}
& a_2b_3 - a_3b_2 \\
& a_3b_1 - a_1b_3 \\
& a_1b_2 - a_2b_1
\end{aligned}
$$

Vector multiplication is useful to find a direction perpendicular to two vectors a and b.

In many cases we need an operation that will turn a vector "a" into another vector "b" with "b" not necessarily along "a." Such is the case when relating a displacement current D to the electric field E causing it. Both E and D have direction and magnitude, hence are vectors. In general D does not lie exactly along E. If D and E are written as vectors on a coordinate system as $E = [E_1, E_2, E_3]$ and $D = [D_1, D_2, D_3]$, we see that a very general linear relation can make each D component proportional to each E component as

$$
\begin{pmatrix} D_1 \\ D_2 \\ D_3 \end{pmatrix} = \begin{pmatrix} \varepsilon_{11}E_1 + \varepsilon_{12}E_2 + \varepsilon_{13}E_3 \\ \varepsilon_{21}E_1 + \varepsilon_{22}E_2 + \varepsilon_{23}E_3 \\ \varepsilon_{31}E_1 + \varepsilon_{32}E_2 + \varepsilon_{33}E_3 \end{pmatrix}
$$

By thinking of the array $\begin{pmatrix} \varepsilon_{11}\varepsilon_{12}\varepsilon_{13} \\ \varepsilon_{21}\varepsilon_{22}\varepsilon_{23} \\ \varepsilon_{31}\varepsilon_{32}\varepsilon_{33} \end{pmatrix}$ as a single entity even if it is multidimensional, we can write $D = \varepsilon E$, which indicates proportionality of D and E, ε being the "constant of proportionality." This multidimensional constant of proportionality is called a matrix, and matrix multiplication expands

$$
\begin{pmatrix} \varepsilon_{11}\varepsilon_{12}\varepsilon_{13} \\ \varepsilon_{21}\varepsilon_{22}\varepsilon_{23} \\ \varepsilon_{31}\varepsilon_{32}\varepsilon_{33} \end{pmatrix} \begin{pmatrix} E_1 \\ E_2 \\ E_3 \end{pmatrix} \quad \text{into} \quad \begin{pmatrix} \varepsilon_{11}E_1 + \varepsilon_{12}E_2 + \varepsilon_{13}E_3 \\ \varepsilon_{21}E_1 + \varepsilon_{22}E_2 + \varepsilon_{23}E_3 \\ \varepsilon_{31}E_1 + \varepsilon_{32}E_2 + \varepsilon_{33}E_3 \end{pmatrix}
$$

Here row is multiplied by column. Using this idea, a scalar product of two vectors a and b is thought of as $(a_1a_2a_3)\begin{pmatrix} b_1 \\ b_2 \\ b_3 \end{pmatrix}$ and $(a_1a_2a_3)$ is thought of as vector "a" transposed. Since we often must transpose matrices, we use a special notation for this:

$$\tilde{a} = (a_1 a_2 a_3) \quad \text{and} \quad \tilde{\varepsilon} = \begin{pmatrix} \varepsilon_{11} & \varepsilon_{21} & \varepsilon_{31} \\ \varepsilon_{12} & \varepsilon_{22} & \varepsilon_{32} \\ \varepsilon_{13} & \varepsilon_{23} & \varepsilon_{33} \end{pmatrix}$$

That is, "rows become columns and columns become rows."
In these terms we can think of the $a \times$ of $a \times b$ as the operator

$$a \times = \begin{pmatrix} 0 & -a_3 & a_2 \\ a_3 & 0 & -a_1 \\ -a_2 & a_1 & 0 \end{pmatrix}$$

THE PIEZOELECTRIC EFFECT AND THE POLARITY TESTER

To guide in interpretation of polarity tests such as those made with Heintzen's polarity tester (Fig. 1.43), we need to consider the effect of crystal symmetry on the piezoelectric effect.

Nonconducting crystals that lack a center of symmetry generally become electrically polarized if stressed in special ways. For example, consider a quartz cube with faces x, y, z. Pressure applied to the x face will cause charges of opposite sign to appear on the x face and on the $-x$ face. The charge is not very large and can leak away in a few seconds if the stress is maintained at a constant value. The charge is proportional to the stress, but stress has six components—tensions along x, y, z, and shearing stresses in the yz, the zx, and the xy planes. If these six components are assigned numbers 1 to 6 so that a stress is given as $T = [T_1 T_2 T_3 T_4 T_5 T_6]$ and the charges on the faces are similarly numbered $1, 2, 3$ for electric displacements parallel to x, y, z, respectively, each of the three components of electric displacement can be written proportional to all six stress components:

$$D_1 = d_{11}T_1 + d_{12}T_2 + d_{13}T_3 + d_{14}T_4 + d_{15}T_5 + d_{16}T_6$$
$$D_2 = d_{21}T_1 + d_{22}T_2 + d_{23}T_3 + d_{24}T_4 + d_{25}T_5 + d_{26}T_6$$
$$D_3 = d_{31}T_1 + d_{32}T_2 + d_{33}T_3 + d_{34}T_4 + d_{35}T_5 + d_{36}T_6$$

or in simple matrix terms, $D = dT$.

Because of symmetry, many of the d terms are necessarily zero and some are equal to others or to others reversed. As a result, for the 32 crystal classes the d matrix arrays are as follows:

Class C_1-1:
$$d = \begin{pmatrix} d_{11} \cdots d_{16} \\ d_{21} \cdots d_{26} \\ d_{31} \cdots d_{36} \end{pmatrix}$$

Class C_i-$\bar{1}$: All d's vanish because of a center of symmetry.

Class C_2-2
(y is binary axis):

$$d = \begin{pmatrix} 0 & 0 & 0 & d_{14} & 0 & d_{16} \\ d_{21} & d_{22} & d_{23} & 0 & d_{25} & 0 \\ 0 & 0 & 0 & d_{34} & 0 & d_{36} \end{pmatrix}$$

Class C_s-m
(y $\perp$ plane of symmetry): •

$$d = \begin{pmatrix} d_{11} & d_{12} & d_{13} & 0 & d_{15} & 0 \\ 0 & 0 & 0 & d_{24} & 0 & d_{26} \\ d_{31} & d_{32} & d_{33} & 0 & d_{35} & 0 \end{pmatrix}$$

C_{2h}-$2/m$. All d's vanish because of center of symmetry.

D_2-222
(x, y, z are binary):

$$d = \begin{pmatrix} 0 & 0 & 0 & d_{14} & 0 & 0 \\ 0 & 0 & 0 & 0 & d_{25} & 0 \\ 0 & 0 & 0 & 0 & 0 & d_{36} \end{pmatrix}$$

C_{2v}-$2mm$
(z is binary, plane of
symmetry $\perp$ x and y):

$$d = \begin{pmatrix} 0 & 0 & 0 & 0 & d_{15} & 0 \\ 0 & 0 & 0 & d_{24} & 0 & 0 \\ d_{31} & d_{32} & d_{33} & 0 & 0 & 0 \end{pmatrix}$$

Class D_{2h}-mm. All d's are zero because of center of symmetry.

Class s_4-$\bar{4}$
(z is $\bar{4}$ axis):

$$d = \begin{pmatrix} 0 & 0 & 0 & d_{14} & d_{15} & 0 \\ 0 & 0 & 0 & -d_{15} & d_{14} & 0 \\ d_{31} & -d_{31} & 0 & 0 & 0 & d_{36} \end{pmatrix}$$

Class C_4-4
(z is 4-fold):

$$d = \begin{pmatrix} 0 & 0 & 0 & d_{14} & d_{15} & 0 \\ 0 & 0 & 0 & d_{15} & -d_{14} & 0 \\ d_{31} & d_{31} & d_{33} & 0 & 0 & 0 \end{pmatrix}$$

Class D_{2d}-$\bar{4}2m$
(z is $\bar{4}$-fold;
x and y binary):

$$d = \begin{pmatrix} 0 & 0 & 0 & d_{14} & 0 & 0 \\ 0 & 0 & 0 & 0 & d_{14} & 0 \\ 0 & 0 & 0 & 0 & 0 & d_{36} \end{pmatrix}$$

Class D_4-422
(x and y binary;
z is 4-fold):

$$d = \begin{pmatrix} 0 & 0 & 0 & d_{14} & 0 & 0 \\ 0 & 0 & 0 & 0 & -d_{14} & 0 \\ 0 & 0 & 0 & 0 & 0 & 0 \end{pmatrix}$$

Class C_{4h}-$4/m$. All d's are zero because of center of symmetry.

C_{4v}-$4mm$
(z is 4-fold;
xz and yz symmetry planes):

$$d = \begin{pmatrix} 0 & 0 & 0 & 0 & d_{15} & 0 \\ 0 & 0 & 0 & d_{15} & 0 & 0 \\ d_{31} & d_{31} & d_{33} & 0 & 0 & 0 \end{pmatrix}$$

Class D_{4h}-4/mmm. All d's are zero because of center of symmetry.

Class C_3-3
(z is trigonal):

$$d = \begin{pmatrix} d_{11} & -d_{11} & 0 & d_{14} & d_{15} & -2d_{22} \\ -d_{22} & d_{22} & 0 & d_{15} & -d_{14} & -2d_{11} \\ d_{31} & d_{31} & d_{33} & 0 & 0 & 0 \end{pmatrix}$$

Class C_{3i}-$\bar{3}$. All d's are zero because of center of symmetry.

Class D_3-32
(z is trigonal;
x binary):

$$d = \begin{pmatrix} d_{11} & -d_{11} & 0 & d_{14} & 0 & 0 \\ 0 & 0 & 0 & 0 & -d_{14} & -2d_{11} \\ 0 & 0 & 0 & 0 & 0 & 0 \end{pmatrix}$$

Class C_{2h}-$\bar{6}$
(z is trigonal axis
mirror plane $\perp z$):

$$d = \begin{pmatrix} d_{11} & -d_{11} & 0 & 0 & 0 & -2d_{22} \\ -d_{22} & d_{22} & 0 & 0 & 0 & -2d_{11} \\ 0 & 0 & 0 & 0 & 0 & 0 \end{pmatrix}$$

Class C_{3v}-3m
(z is trigonal;
yz plane is mirror plane):

$$d = \begin{pmatrix} 0 & 0 & 0 & 0 & d_{15} & -2d_{22} \\ -d_{22} & d_{22} & 0 & d_{15} & 0 & 0 \\ d_{31} & d_{31} & d_{33} & 0 & 0 & 0 \end{pmatrix}$$

Class D_{3d}-$\bar{3}m$. All d's vanish because of center of symmetry.

Class D_{3h}-$\bar{6}m2$
(z is trigonal;
xy and zx planes
are mirror planes):

$$d = \begin{pmatrix} d_{11} & -d_{11} & 0 & 0 & 0 & 0 \\ 0 & 0 & 0 & 0 & 0 & -2d_{11} \\ 0 & 0 & 0 & 0 & 0 & 0 \end{pmatrix}$$

Class C_6-6
(z is hexagonal):

$$d = \begin{pmatrix} 0 & 0 & 0 & d_{14} & d_{15} & 0 \\ 0 & 0 & 0 & d_{15} & -d_{14} & 0 \\ d_{31} & d_{31} & d_{33} & 0 & 0 & 0 \end{pmatrix}$$

Class D_6-622
(z is hexagonal;
x and y binary):

$$d = \begin{pmatrix} 0 & 0 & 0 & d_{14} & 0 & 0 \\ 0 & 0 & 0 & 0 & -d_{14} & 0 \\ 0 & 0 & 0 & 0 & 0 & 0 \end{pmatrix}$$

Class C_{6h}-6/m. All d's vanish because of center of symmetry.

Class C_{6h}-6mm
(z is hexagonal;
ax and zy mirror
planes):

$$d = \begin{pmatrix} 0 & 0 & 0 & 0 & d_{15} & 0 \\ 0 & 0 & 0 & d_{15} & 0 & 0 \\ d_{31} & d_{31} & d_{33} & 0 & 0 & 0 \end{pmatrix}$$

Class D_{6h}-6/mmm. All d's vanish because of center of symmetry.

Class T-23
(x, y, and z are binary):

$$d = \begin{pmatrix} 0 & 0 & 0 & d_{14} & 0 & 0 \\ 0 & 0 & 0 & 0 & d_{14} & 0 \\ 0 & 0 & 0 & 0 & 0 & d_{14} \end{pmatrix}$$

Class $T_d = \bar{4}3m$. Same as class T-23.

Class O-432. All d's are zero, no center of symmetry but too many other symmetries.

Class T_h-$m3$. All d's vanish because of center of symmetry.

Class T_d-$\bar{4}3m$
(x, y, and z are 4-fold):

$$d = \begin{pmatrix} 0 & 0 & 0 & d_{14} & 0 & 0 \\ 0 & 0 & 0 & 0 & d_{14} & 0 \\ 0 & 0 & 0 & 0 & 0 & d_{14} \end{pmatrix}$$

Class O_h-$m3m$. All d's vanish because of center of symmetry.

We now illustrate the use of the equation $D = dT$ to calculate the displacement current caused by pressing on the x face of a quartz cube.

$$D = \begin{pmatrix} d_{11} & -d_{11} & 0 & d_{14} & 0 & 0 \\ 0 & 0 & 0 & 0 & -d_{14} & -2d_{11} \\ 0 & 0 & 0 & 0 & 0 & 0 \end{pmatrix} \begin{pmatrix} 1 \\ 0 \\ 0 \\ 0 \\ 0 \\ 0 \end{pmatrix} = \begin{pmatrix} d_{11} \\ 0 \\ 0 \end{pmatrix}$$

This tells us that there will be an accumulation of charge on the x and $-x$ faces. If we pressed on a y face we would have $T = [010000]$, giving

$$D = dT = \begin{pmatrix} -d_{11} \\ 0 \\ 0 \end{pmatrix}$$; that is, pressing on y faces gives an accumulation of

charge on x faces.

The piezo effect is reversible in the sense that an electric field causes a strain in the crystal and the strain-field relation is simply the d matrix transposed. In this way we calculate the strain on the above-mentioned quartz cube, the strain being caused by applying an electric field E_1 along the x axis:

$$S = E = \begin{pmatrix} d_{11} & 0 & 0 \\ -d_{11} & 0 & 0 \\ 0 & 0 & 0 \\ d_{14} & 0 & 0 \\ 0 & -d_{14} & 0 \\ 0 & -2d_{11} & 0 \end{pmatrix} \begin{pmatrix} E_1 \\ 0 \\ 0 \end{pmatrix} = \begin{pmatrix} d_{11}E_1 \\ -d_{11}E_1 \\ 0 \\ d_{14}E_1 \\ 0 \\ 0 \end{pmatrix}$$

We now know that the cube will expand along the x direction, contract along the y direction, and also shear about the x axis (i.e., in the yz plane). At least this is the case if d_{11} is positive and $+x$ is made positive.

Let us now examine a cube of ammonium dihydrogen phosphate (ADP). The cube is to have x faces, y faces, and z faces. A pressure on the x faces causes a displacement current

$$
\begin{pmatrix}
0 & 0 & 0 & d_{14} & 0 & 0 \\
0 & 0 & 0 & 0 & d_{14} & 0 \\
0 & 0 & 0 & 0 & 0 & d_{36}
\end{pmatrix}
\begin{pmatrix}
T_1 \\ 0 \\ 0 \\ 0 \\ 0 \\ 0
\end{pmatrix}
=
\begin{pmatrix}
0 \\ 0 \\ 0
\end{pmatrix}
$$

Similarly a pressure on the y faces gives [000] and a pressure on the z faces gives $D = [000]$. That is, the polarity tester of Fig. 1.43 can find no polarity along any direction caused by pressure on any of the cube faces. However a shearing stress in the yz plane $T = [000T_400]$ gives $D = [d_{14}T_4,$ 0, 0], a charge on the x faces. Also a shearing stress in the zx plane $T = [0000T_50]$ gives $D = [0, d_{14}T_5, 0]$, and finally a shearing stress in the xy plane $T = [00000T]$ gives $D = [0, 0, d_{36}T_6]$. The yz shear can be approximated by compression or tension along the direction $y = z$ or $y = -z$ to give the charges on the x faces. Similarly stress along the direction $z = \pm x$ gives charges on the y faces and stress along the direction $y = \pm x$ gives charges on the z faces. This may be clearer if we choose new x and y axes at 45° to the old ones but still 90° from z. In this case the d matrix in terms of the old d values is

$$
d' =
\begin{pmatrix}
0 & 0 & 0 & 0 & d_{14} & 0 \\
0 & 0 & 0 & -d_{14} & 0 & 0 \\
d_{36/2} & -d_{36/2} & 0 & 0 & 0 & 0
\end{pmatrix}
$$

On these new axes $T' = [T_1'00000]$ gives $D' = [0, 0, d_{36}T_1'/2]$, that is, a polarization along z. Similarly a stress $T' = [0, T_2'0000]$ gives $D' = [0, 0, -d_{36}T_2'/2]$, again, charges on the z faces.

Thus we see that most piezoelectric crystals can develop a polarization along the direction of simple tension or compression if this direction is carefully chosen. Only in classes $D_4 - 422$ and $D_6 - 622$ is this untrue.

SOME SAMPLE OPTICAL CALCULATIONS FOR AN ORTHORHOMBIC CRYSTAL

Argonite has $a_0 = 4.94$ Å, $b_0 = 7.94$ Å, $C_0 = 5.72$ Å, $X = c$, $Y = a$, $Z = b$, $N_\alpha = 1.531$, $N_\beta = 1.682$, $N_\gamma = 1.686$. Hence

$$g = \begin{pmatrix} N_\alpha^{-2} & 0 & 0 \\ 0^\beta & N_\beta^{-2} & 0 \\ 0 & 0^\gamma & N_\gamma^{-2} \end{pmatrix} = \begin{pmatrix} 0.3535 & 0 & 0 \\ 0 & 0.3518 & 0 \\ 0 & 0 & 0.4266 \end{pmatrix}$$

A light ray enters a (111) face perpendicularly. What are the polarization directions inside the crystal and the refractive indices for these?

Here the wave normal is $[1/a_0, 1/b_0, 1/c_0] = [0.2024, 0.1259, 0.1748]$, which normalizes to the unit vector $[0.6847, 0.4260, 0.5913] = s$. With this wave normal the two unit displacement vectors are

$$D' = u \cos \theta + v \sin \theta$$
$$D'' = u \sin \theta - v \cos \theta$$

where u and v are two arbitrary mutually perpendicular unit vectors both perpendicular to vector s, and θ is an angle such that $\tan 2\theta = 2(\bar{u}gv)/(\bar{u}gu - \bar{v}gv)$, which expands to

$$\tan 2\theta = \frac{2(u_1 v_1 g_{11} + u_2 v_2 g_{22} + u_3 v_3 g_{33})}{(u_1^2 - v_1^2)g_{11} + (u_2^2 - v_2^2)g_{22}(u_3^2 - v_3^2)g_{33}}$$

We take u along $[001] \times s$ (i.e., along $[-0.4260, 0.6847, 0]$), which normalizes to $u = [-0.5282, 0.8491, 0]$; v is taken as $u \times s = [0.5021, 0.3123, -0.8064] = v$, whence $\tan 2\theta = 1.044°$, $\theta = 0.522°$, $D' = [-0.5237, 0.8519, 0.0073]$, $D'' = [0.5073, 0.3046, 0.8064]$, where all signs have been reversed for convenience, since this makes no difference in actual use.

For a current displacement along unit vector D the refractive index n is given by $n^{-2} = \bar{D}gD$, which expands to

$$n^{-2} = D_1^2 g_{11} + D_2^2 g_{22} + D_3^2 g_{33}$$

From this, for D', we calculate $(n')^{-2} = 0.3523$, $n' = 1.685$ $(D'')^{-2} = 0.4010$, $n'' = 1.579$.

CRYSTAL ELASTICITY

To show how acoustic velocities vary with direction, we need to introduce the elastic moduli that relate stress and strain. We have seen that the specification of a stress T requires the statement of the values of six components, T_1, T_2, T_3, T_4, T_5, T_6. Here T_1 is the x component of the force per unit area over an area perpendicular to the x axis, T_2 is the y component of the force per unit area over an area perpendicular to the y axis, and similarly T_3 is the z component of the force per unit area over an area perpendicular to the z axis. Components T_4, T_5, and T_6 represent tangential forces, that is, shearing forces—T_4 in the yz plane, T_5 in the zx plane, and T_6 in the xy plane.

In a parallel manner we list six components of strain, $S_1, S_2, \ldots, S_6$. Here S_1 represents fractional elongation parallel to the x axis, S_2 represents fractional elongation parallel to the y axis, and similarly for S_3. Also S_4, S_5, and S_6 represent shear in the yz plane, zx plane, and the xy plane, respectively.

Within the elastic limits, the stresses are proportional to the strains. This can be written

$$T_1 = C_{11}S_1 + C_{12}S_2 + C_{13}S_3 + C_{14}S_4 + C_{15}S_5 + C_{16}S_6$$
$$T_2 = C_{21}S_1 \ldots$$
$$\cdots\cdots\cdots\cdots\cdots\cdots\cdots$$
$$T_6 = C_{61}S_1 \cdots\cdots\cdots\cdots\cdots\cdots\cdots\cdots\cdots\cdots + C_{66}S_6$$

Or, in matrix notation, $T = CS$, where the 36 quantities C_{ij} are the elastic moduli. It can be shown that always $C_{ij} = C_{ji}$. This reduces the number of moduli for the least symmetric crystal class to 21. Crystals of a higher symmetry need fewer moduli, as shown below.

Classes C_1-1 and C_i-1. (21 moduli):

$$C = \begin{pmatrix} C_{11} & C_{12} & C_{13} & C_{14} & C_{15} & C_{16} \\ C_{12} & C_{22} & C_{23} & C_{24} & C_{25} & C_{26} \\ C_{13} & C_{23} & C_{33} & C_{34} & C_{35} & C_{36} \\ C_{14} & C_{24} & C_{34} & C_{44} & C_{45} & C_{46} \\ C_{15} & C_{25} & C_{35} & C_{45} & C_{55} & C_{56} \\ C_{16} & C_{26} & C_{36} & C_{46} & C_{56} & C_{66} \end{pmatrix}$$

Classes C_2-2, C_s-m, and C_{2h}-$2/m$. (13 moduli) (y is the unique axis):

$$C = \begin{pmatrix} C_{11} & C_{12} & C_{13} & 0 & C_{15} & 0 \\ C_{12} & C_{22} & C_{23} & 0 & C_{25} & 0 \\ C_{13} & C_{23} & C_{33} & 0 & C_{35} & 0 \\ 0 & 0 & 0 & C_{44} & 0 & C_{46} \\ C_{15} & C_{25} & C_{35} & 0 & C_{55} & 0 \\ 0 & 0 & 0 & C_{46} & 0 & C_{66} \end{pmatrix}$$

Classes D_2-222, C_{2v}-$2mm$, and D_{2h}-mmm. (9 moduli):

$$C = \begin{pmatrix} C_{11} & C_{12} & C_{13} & 0 & 0 & 0 \\ C_{12} & C_{22} & C_{23} & 0 & 0 & 0 \\ C_{13} & C_{23} & C_{33} & 0 & 0 & 0 \\ 0 & 0 & 0 & C_{44} & 0 & 0 \\ 0 & 0 & 0 & 0 & C_{55} & 0 \\ 0 & 0 & 0 & 0 & 0 & C_{66} \end{pmatrix}$$

Classes S_4-$\bar{4}$, C_4-4, and C_{4h}-$4/m$. (7 moluli):

$$C = \begin{pmatrix} C_{11} & C_{12} & C_{13} & 0 & 0 & C_{16} \\ C_{12} & C_{11} & C_{13} & 0 & 0 & C_{16} \\ C_{13} & C_{13} & C_{33} & 0 & 0 & 0 \\ 0 & 0 & 0 & C_{44} & 0 & 0 \\ 0 & 0 & 0 & 0 & C_{44} & 0 \\ C_{16} & C_{16} & 0 & 0 & 0 & C_{66} \end{pmatrix}$$

Classes D_{2d}-$\bar{4}2m$, D_4-422, C_{4v}-$4mm$, and D_{4h}-$4/mmm$. (6 moduli) (z axis is fourfold):

$$C = \begin{pmatrix} C_{11} & C_{12} & C_{13} & 0 & 0 & 0 \\ C_{12} & C_{11} & C_{13} & 0 & 0 & 0 \\ C_{13} & C_{13} & C_{33} & 0 & 0 & 0 \\ 0 & 0 & 0 & C_{44} & 0 & 0 \\ 0 & 0 & 0 & 0 & C_{44} & 0 \\ 0 & 0 & 0 & 0 & 0 & C_{66} \end{pmatrix}$$

Classes C_3-3 and C_{3i}-$\bar{3}$. (7 moduli) (z is trigonal axis):

$$C = \begin{pmatrix} C_{11} & C_{12} & C_{13} & C_{14} & -C_{25} & 0 \\ C_{12} & C_{11} & C_{13} & -C_{14} & C_{25} & 0 \\ C_{13} & C_{13} & C_{33} & 0 & 0 & 0 \\ C_{14} & -C_{14} & 0 & C_{44} & 0 & C_{25} \\ -C_{25} & C_{25} & 0 & 0 & C_{44} & C_{14} \\ 0 & 0 & 0 & C_{25} & C & \dfrac{C_{11}-C_{12}}{2} \end{pmatrix}$$

Classes D_3-32, C_{3v}-$3m$, C_{2i}-$\bar{3}$, and D_{3d}-$\bar{3}m$. (6 moduli) (z is trigonal, x is binary, or $y \perp$ to mirror plane):

$$
C = \begin{pmatrix}
C_{11} & C_{12} & C_{13} & C_{14} & 0 & 0 \\
C_{12} & C_{11} & C_{13} & -C_{14} & 0 & 0 \\
C_{13} & C_{13} & C_{33} & 0 & 0 & 0 \\
C_{14} & -C_{14} & 0 & C_{44} & 0 & 0 \\
0 & 0 & 0 & 0 & C_{44} & C_{14} \\
0 & 0 & 0 & 0 & C_{14} & \dfrac{C_{11}-C_{12}}{2}
\end{pmatrix}
$$

Classes C_{3h}-$\bar{6}$, D_{3h}-$\bar{6}m2$, C_6-6, D_6-622, C_{6h}-$6/m$, C_{6v}-$6mm$, and D_{6h}-$6/mmm$. (5 moduli) (z is hexagonal):

$$
C = \begin{pmatrix}
C_{11} & C_{12} & C_{13} & 0 & 0 & 0 \\
C_{12} & C_{11} & C_{13} & 0 & 0 & 0 \\
C_{13} & C_{13} & C_{33} & 0 & 0 & 0 \\
0 & 0 & 0 & C_{44} & 0 & 0 \\
0 & 0 & 0 & 0 & C_{44} & 0 \\
0 & 0 & 0 & 0 & 0 & \dfrac{C_{11}-C_{12}}{2}
\end{pmatrix}
$$

Classes T-23, O-432, T_h-$m3$, T_d-$\bar{4}3m$, and O_h-$m3m$. (cubic crystals using cubic axes; 3 moduli):

$$
C = \begin{pmatrix}
C_{11} & C_{12} & C_{12} & 0 & 0 & 0 \\
C_{12} & C_{11} & C_{12} & 0 & 0 & 0 \\
C_{12} & C_{12} & C_{11} & 0 & 0 & 0 \\
0 & 0 & 0 & C_{44} & 0 & 0 \\
0 & 0 & 0 & 0 & C_{44} & 0 \\
0 & 0 & 0 & 0 & 0 & C_{44}
\end{pmatrix}
$$

For isotropic bodies the matrix is the same except that $C_{44} = (C_{11} - C_{12})/2$, that is, two moduli.

To illustrate the use of the matrices just given, we consider a unit cube of silicon having faces (100), (010), and (001). For silicon, $C_{11} = 16.740 \times 10^{11}$ dynes/cm^2, $C_{12} = 6.532 \times 10^{11}$ dynes/cm^2, and $C_{44} = 7.957 \times 10^{11}$ dynes/cm^2. To cause the z dimension to shrink by one part in 10^4, that is, $S = [0, 0 - 10^{-4}, 0, 0, 0]$, we compute the necessary stress as $T = CS$ or $T = [C_{12}S_3, C_{12}S_3, C_{11}S_3, 0, 0, 0]$. Hence we must apply a pressure of 6.523×10^7 dynes/cm^2 on the x and y faces and a pressure of 16.74×10^7 dynes/cm^2 on the z faces. This is 66 kg on the x and y faces,

170 kg on the z faces. If we apply only the 170 kg to the z faces, the crystal will expand in the x and y directions, but our strain specification was for no change in the x or y direction.

Calculation of Velocity of Elastic Waves in Crystals

The problem of elastic waves in crystals is more complicated than the problem of light waves in crystals. The most general crystal, a triclinic crystal, has 21 independent elastic constants. Isotropic substances have only two independent elastic constants, but even cubic crystals have three independent constants.

In light waves, for any direction of a wave normal there are two waves possible, two velocities of propagation, and two mutually perpendicular directions of polarization.

In elastic waves there are three waves, three velocities, and three mutually perpendicular "polarizations" (i.e., particle motions). The solution of a cubic equation is generally required to find those polarizations and velocities.

To solve for the velocities along the unit vector wave normal $[l, m, n]$ we start with the universal symbolic pseudo-matrix:

$$\bar{C} = \begin{pmatrix} C_1 & C_6 & C_5 \\ C_6 & C_2 & C_4 \\ C_5 & C_4 & C_3 \end{pmatrix}$$

we multiply $\bar{C}$ by the vector $[lmn]$ to get a pseudo-vector φ:

$$\varphi = \bar{C} \begin{pmatrix} l \\ m \\ n \end{pmatrix}$$

We then multiply φ by its transpose. In the product, a 3×3 matrix, we replace each $C_i C_j$ by C_{ij}, where the C_{ij}'s are the elastic moduli. Thus written, the resulting matrix is the 3×3 dynamic matrix Γ. There are three "eigenvalues" $\Lambda_{(i)}$ for this matrix and three "eigenvectors" $D_{(i)}$, each obeying the equation $\Gamma D_{(i)} = \Lambda_{(i)} D_{(i)}$. Here each of the three values of $\Lambda_{(i)}$ is a value of density times wave velocity squared (ρv_i^2) and each of the three vectors $D_{(i)}$ is the corresponding direction of particle motion for its wave. The equation for $\Lambda_{(i)}$ can be solved by solving the determinantal equation

$$\begin{vmatrix} \Gamma_{11} - \Lambda_{(i)} & \Gamma_{12} & \Gamma_{13} \\ \Gamma_{12} & \Gamma_{22} - \Lambda_{(i)} & \Gamma_{23} \\ \Gamma_{13} & \Gamma_{23} & \Gamma_{33} - \Lambda_{(i)} \end{vmatrix} = 0$$

which is the cubic equation $\Lambda^3 - T\Lambda^2 + S\Lambda = \Delta$, where T is the "trace" (i.e., the sum of the major diagonal terms of the matrix Γ), S is the sum of the minors of the major diagonal terms, and Δ is the value of the determinant $|\Gamma|$.

If each major diagonal term Γ_{11}, Γ_{22}, Γ_{33} is reduced by a quantity Q, a new dynamic matrix Γ' is formed and the three eigenvalues of Γ' are smaller than those of Γ by the amount Q. In particular if we make Q equal to one-third of $\Gamma_{11} + \Gamma_{22} + \Gamma_{33}$, the matrix Γ' has its trace equal to zero and the sum of its eigenvalues is zero. Hence the quadratic term vanishes. If then we write $y_{(i)}$ for $\Lambda_{(i)} - Q$ we have $y^3 + S'y = \Delta'$, where Δ' is the value of $|\Gamma'|$ and S' is the sum of the minors of the major diagonal of Γ'. Here S' is always negative and we let $-S' = \sigma$, a positive number. The equation for y is now $y^3 - \sigma y - \Delta' = 0$. We take $\cos\alpha = \dfrac{\Delta'}{2}\sqrt{27/\sigma^3}$ and write k for $2\sqrt{\sigma/3}$. The three solutions for y are

$$y_{(1)} = k\,\cos\left(\frac{\alpha}{3}\right)$$

$$y_{(2)} = -k\,\sin\left(\frac{\alpha}{3} + 30°\right)$$

$$y_{(3)} = k\,\sin\left(\frac{\alpha}{3} - 30°\right)$$

To complete the calculation for the eigenvalues of Γ we must now remember that $\Lambda_{(i)} = y_{(i)} + Q$.

To find the eigenvector (i.e., the direction of the particle motion) for any eigenvalue $\Lambda_{(i)}$, we subtract $\Lambda_{(i)}$ from each term Γ_{11}, Γ_{22}, Γ_{33}; then we find the three cofactors of any row of the resulting matrix.

We will take as an example, silicon a cubic crystal. Hence the elastic modulus matrix is

$$C = \begin{pmatrix} C_{11} & C_{12} & C_{12} & 0 & 0 & 0 \\ C_{12} & C_{11} & C_{12} & 0 & 0 & 0 \\ C_{12} & C_{12} & C_{11} & 0 & 0 & 0 \\ 0 & 0 & 0 & C_{44} & 0 & 0 \\ 0 & 0 & 0 & 0 & C_{44} & 0 \\ 0 & 0 & 0 & 0 & 0 & C_{44} \end{pmatrix}$$

which tells us which C_{ij}'s vanish and which terms are equal. Here, for example, $C_{11} = C_{22} = C_{33} = 16.740 \times 10^{11}$ dynes/cm^2, $C_{12} = C_{13} = C_{23} = 6.523 = 10^{11}$ dynes/cm^2, and $C_{44} = C_{55} = C_{66} = 7.957 \times 10^{11}$ dynes/cm^2. Also $\rho = 2.331$ g/cm^3. For the wave normal $[lmn]$ we find

$$\varphi = \begin{pmatrix} C_1 & C_6 & C_5 \\ C_6 & C_2 & C_4 \\ C_5 & C_4 & C_3 \end{pmatrix}\begin{pmatrix} l \\ m \\ n \end{pmatrix} = \begin{pmatrix} lC_1 + mC_6 + nC_5 \\ lC_6 + mC_2 + nC_4 \\ lC_5 + mC_4 + nC_3 \end{pmatrix}$$

There are three "eigenvectors" $D_{(i)}$, each obeying the equation

$$\Gamma D_{(i)} = \Lambda_{(i)} D_{(i)}$$

from which we obtain

$$\Gamma = \begin{pmatrix} l^2 C_{11} + (m^2 + n^2)C_{44} & ml(C_{12} + C_{44}) & nl(C_{12} + C_{44}) \\ lm(C_{12} + C_{44}) & m^2 C_{11} + (n^2 + l^2)C_{44} & mn(C_{12} + C_{44}) \\ ln(C_{12} + C_{44}) & mn(C_{12} + C_{44}) & n^2 C_{11} + (l^2 + m^2)C_{44} \end{pmatrix}$$

We choose $[lmn]$ as $[0.9000, 0.2500, 0.35707]$, which gives

$$\Gamma = \begin{pmatrix} 15.07123 & 3.2580 & 4.65335 \\ 3.25800 & 8.50594 & 1.29260 \\ 4.65335 & 1.29260 & 9.07683 \end{pmatrix}$$

from which $Q = 10.8847$ and

$$\Gamma' = \begin{pmatrix} 4.1866 & 3.2580 & 4.65335 \\ 3.2580 & -2.3787 & 1.29260 \\ 4.65335 & 1.2926 & -1.80787 \end{pmatrix} \Bigg\} \begin{cases} \sigma = 47.16606 \\ k = 7.9302 \\ \Delta' = 120.8996 \end{cases}$$

$$\cos \alpha = \frac{\Delta'}{2} \sqrt{27/\sigma^3} = 0.96969, \quad \alpha = 14.142656°, \quad \frac{\alpha}{3} = 4.71422°$$

$$y_{(1)} = k \cos (\alpha/3)$$
$$y_{(2)} = -k \sin (\alpha/3 + 30°)$$
$$y_{(3)} = k \sin (\alpha/3 - 30°)$$

$\Lambda_{(1)} = y_{(1)} + Q = 18.7880$, $\Lambda_{(2)} = 6.3685$, $\Lambda_{(3)} = 7.4974$.

From these relations we have $v_{(1)} = (\Lambda_{(1)}/\rho)^{1/2} = (18.7879 \times 10^{11}/2.331)^{1/2} = 8.978 \times 10^5$ cm/sec; similarly $v_{(2)} = 3.162 \times 10^5$ cm/sec, and $v_{(3)} = 5.671 \times 105$ cm/sec. We check that $y_{(1)} + y_{(2)} + y_{(3)} = 0$ (actually 0.00003). We see that $\Lambda_{(1)} + \Lambda_{(2)} + \Lambda_{(3)} = 32.6539$, while $\Gamma_{11} + \Gamma_{22} + \Gamma_{33} = 32.6539$.

Let us subtract $\Lambda_{(1)}$ from the major diagonal terms of Γ:

$$\Gamma_{(1)} = \begin{vmatrix} -3.71678 & 3.2580 & 4.6533 \\ 3.2580 & -10.28207 & 1.2926 \\ 4.65335 & 1.2926 & -9.7112 \end{vmatrix}$$

This determinant has a value of 0.002354 instead of zero because of rounding off errors. The cofactors of the top row of this determinant—98.1804, 37.6540, and 52.0574—give the direction of the particle motion. This vector normalizes to $[0.83676, 0.32091, 0.44367]$, which is almost but not quite along the wave normal $[0.90, 0.25, 0.35707]$. The angle between the motion and the wave normal is 7.36°. The wave is *quasi*-longitudinal.

A further test is that $\Gamma D_{(1)} = \Lambda D_{(1)}$:

$$
\begin{pmatrix} 15.0712 & 3.2580 & 4.6534 \\ 3.2580 & 8.5059 & 1.2926 \\ 4.6534 & 1.2926 & 9.0768 \end{pmatrix} \begin{pmatrix} 0.83676 \\ 0.32091 \\ 0.44367 \end{pmatrix} = \begin{pmatrix} 15.73208 \\ 6.02928 \\ 8.33569 \end{pmatrix} = 18.788 \begin{pmatrix} 0.83676 \\ 0.32091 \\ 0.44367 \end{pmatrix}
$$

If we subtract $\Lambda_{(2)}$ from the major diagonal elements of Γ, we have

$$
\Gamma_{(2)} = \begin{pmatrix} 8.7027 & 3.2580 & 4.6533 \\ 3.2580 & 2.1374 & 1.2926 \\ 4.6533 & 1.2926 & 2.7083 \end{pmatrix}
$$

Cofactors for the first row are 4.1179, -2.8088, -5.7347, which normalizes to the unit vector [0.5420, -0.3697, -0.7547]. Taking the scalar product of this with the wave normal unit vector shows the angle between them to be 82.77°. To be truly a transverse (or shear) wave, this angle should be 90°. We call it a quasi-transverse wave. Checking $\Gamma D = \Lambda D$, we find $\Gamma D_{(2)} = 6.3686$ [0.5421, -0.3697, -0.7546], while $\Lambda D_{(3)} = 6.3685$ [0.5420, -0.3697, -0.7546]. Finally, subtracting $\Lambda_{(3)}$ from the major diagonal element of Γ gives

$$
\Gamma_{(3)} = \begin{pmatrix} 7.5738 & 3.2580 & 4.6533 \\ 3.2580 & 1.0085 & 1.2926 \\ 4.6533 & 1.2926 & 1.5794 \end{pmatrix}
$$

The cofactors of the second row are -0.86917, 9.69114, -5.37056, which normalizes to the unit vector [0.0782, -0.8720, 0.4832]. This vector makes an angle of 88.57° from the wave normal, and $\Gamma D_{(3)} = \Lambda_{(3)} D_{(3)}$ quite accurately.

Checking the scalar products of $D_{(1)}$, $D_{(2)}$, and $D_{(3)}$, we find that the angle between $D_{(1)}$ and $D_{(2)}$ is 89.997°, the angle between $D_{(2)}$ and $D_{(3)}$ is 89.995°, and the angle between $D_{(3)}$ and $D_{(1)}$ is 90.001°. Departures from exactly 90° are due to rounding off errors.

To recapitulate, an acoustic wave traveling along [0.9000, 0.2500, 0.35707] can have a quasi-longitudinal wave with particle motion along [0.83676, 0.32091, 0.44367] and a normal wave velocity of 8.978 km/sec, a quasi-transverse wave with particle motion along [0.5420, -0.3697, -0.7546] and normal velocity 3.162 km/sec, and also another quasi-transverse wave with particle motion along [0.0782, -0.8720, 0.4832] traveling with a normal velocity of 5.671 km/sec. There can be any combination of these with arbitrary relative amplitudes.

To illustrate the dependence of acoustic velocities on direction we choose a simple example, the velocity-direction dependence for silicon in the (001) plane. Here

$$\varphi = \begin{pmatrix} C_1 & C_6 & C_5 \\ C_6 & C_2 & C_4 \\ C_5 & C_4 & C_3 \end{pmatrix} \begin{pmatrix} \cos \alpha \\ \sin \alpha \\ 0 \end{pmatrix} = \begin{pmatrix} C_1 \cos \alpha + C_6 \sin \alpha \\ C_6 \cos \alpha + C_2 \sin \alpha \\ C_5 \cos \alpha + C_4 \sin \alpha \end{pmatrix}$$

Making use of the fact that $C_i C_j$ becomes C_{ij} and recalling that for cubic crystals C_{16}, C_{26}, C_{15}, C_{25}, C_{24}, C_{56}, and C_{46} are all zero and $C_{44} = C_{55} = C_{66}$, we find

$$\varphi\tilde{\varphi} = \Gamma = \begin{pmatrix} C_{11} \cos^2 \alpha + C_{44} \sin^2 \alpha & (C_{12} + C_{44}) \sin \alpha \cos \alpha & 0 \\ (C_{12} + C_{44}) \sin \alpha \cos \alpha & C_{11} \sin^2 \alpha + C_{44} \cos^2 \alpha & 0 \\ 0 & 0 & C_{44} \end{pmatrix}$$

Solving the equation

$$\begin{vmatrix} \Gamma_{11} - \Lambda & \Gamma_{12} & 0 \\ \Gamma_{12} & \Gamma_{22} - \Lambda & C \\ 0 & 0 & C_{44} - \Lambda \end{vmatrix} = 0$$

we have one solution immediately: $\Lambda = \rho v^2 = C_{44}$; we also have $\Lambda^2 - (\Gamma_{11} + \Gamma_{22})\Lambda + \Gamma_{11}\Gamma_{22} - \Gamma_{12}^2 = 0$, from which

$$\Lambda = \frac{\Gamma_{11} + \Gamma_{22}}{2} \pm \sqrt{\left(\frac{\Gamma_{11} - \Gamma_{22}}{2}\right)^2 + \Gamma_{12}^2}$$

This expands to

$$\Lambda = \frac{C_{11} + C_{44}}{2} \pm \sqrt{\left(\frac{C_{11} - C_{44}}{2}\right)^2 \cos^2 2\alpha + \left(\frac{C_{12} + C_{44}}{2}\right)^2 \sin^2 2\alpha}$$

For silicon with $C_{11} = 16.740 \times 10^{11}$ dynes/cm^2, $C_{12} = 6.523 \times 10^{11}$ dynes/cm^2, and $C_{44} = 7.957 \times 10^{11}$ dynes/cm^2, this becomes

$$\Lambda = \{12.348 \pm 4.391 \sqrt{1 + 1.718 \times \sin^2 2\alpha}\} \times 10^{11}$$

The zeros in the Γ determinant simplified the solution by avoiding the cubic equation of the more general solution. The results are plotted in Fig. 13.9. The outer curve is for longitudinal waves, the inner curve is for shear waves where the particle motion is shown by the arrows. The center curve, a circle, is a shear wave with particle motion perpendicular to the paper (i.e., the z axis).

X-RAY REFLECTION INTENSITIES

The X-ray reflection intensities quoted in Chapter 6 are mainly from powder data. The data are readily available—thousands of substances are

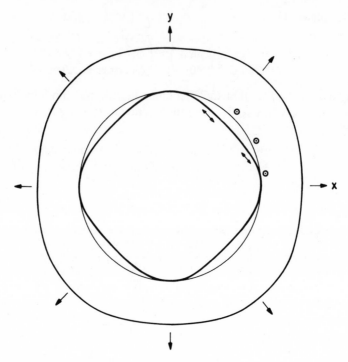

Figure 13.9 Velocity versus direction in silicon.

covered with *hkl* indices, interplanar spacing, and relative intensities. There is an awkward weakness in that sometimes two planes have the same or nearly the same *d* spacing, hence are not resolved by powder methods. For example, since cubic 333 and 511 have the same *d* spacing, the entry 333, 511, *I* = 40 might be found in the tables. In this case the sum of the two is 40 and we have left open the question of dividing the 40 into the part due to 333 and that due to 511.

Let us briefly study the factors that determine these intensities. The main fact is that X-rays diffracted from individual atoms travel in certain directions with more constructive interference than those diffracted in other directions. This is true even for a single atom, because the diffraction is from the electrons outside the nucleus, and these are distributed in the space around the nucleus. For diffraction at small angles the electromagnetic waves from all electrons are essentially in phase, whereas at large angles the waves from regions further from the source are out of phase with waves from regions nearer the source. Hence the results from a single atom depend on the distribution of electrons about that atom. We know that for each kind of atom there are preferred orbits, hence preferred distances from the nucleus center. Some atoms have hundreds of electrons per cubic angstrom, others have but few. An ionized atom differs from the same kind of atom in a neutral state. Quantum mechanics has spelled out the radial distribution of electron densities for all known atoms and has calculated factors that allow for the angle effect at various angles of deflection. These f or atomic scattering factors start with the value Z (the atomic number) for zero deflection and fall as $(\sin \theta)/\lambda$ grows larger. Graphs and tables of these factors are available in crystal diffraction literature.

In general a unit cell contains more than one atom and more than one molecule; to calculate the intensity from the whole cell, we must take all the atomic scattering factors for the angle in question and add them, keeping account of their phases. To do this we consider the jth atom in a cell and calculate a structure factor F for the cell by calculating for each atom the value of $f_j e^{2\pi i(hx_j + ky_j + lz_j)}$. Here the reflection is from the plane (hkl) and x_j, y_j, z_j are the coordinates of the jth atom. These coordinates are reckoned on the unit cell axes, which may or may not be isometric axes. Hence to find the structure factor for the whole cell we add the effects of all the atoms:

$$F = \sum_j f_j e^{2\pi i(hx_j + ky_j + lz_j)}$$

which keeps the phase account for $e^{i\varphi} = \cos \varphi + i \sin \varphi$. This can be written

$$F = \sum_j f_j \cos [(2\pi(hx_j + ky_j + lz_j)]$$
$$+ i \sum_j f_j \sin [2\pi(hx_j + ky_j + lz_j)]$$

which obviously keeps account of the phases between atoms.

For ideally imperfect crystals the intensity of reflection is proportional to F^2. By "ideally imperfect" we mean that there is a small amount of disorder in the assembly of unit cells (i.e., the translations are not ideal). For ideally perfect crystals I is proportional to F.

Measurement of the Bragg angles of several planes (hkl) determine the unit cell without the need to measure intensities. It is the task of those who determine crystal structure to devise an arrangement of the atoms known to be in the unit cell, an arrangement that best fits the observed reflection intensities. Some crystals have all atoms on symmetry planes. For these the values of x, y, z for each atom is at once evident. For many crystals some atoms are determined wholly or partially by symmetry, but some parameters are left to be calculated. For other atoms all parameters x, y, and z must be calculated.

Several factors have been omitted from the foregoing treatment. Temperature causes all atoms to vibrate about their rest positions, smearing out the effects contributing to the structure factors. Also the reflected beam is weakened because the rays travel through other atoms besides the ones we considered as diffracting. There is also a polarization effect. An electromagnetic ray traveling along a certain direction and vibrating along a perpendicular direction but scattering in another direction cannot contribute its entire amplitude to the new direction but only a component of it. If the scattered ray is perpendicular to the incoming ray, this component can be zero. The net result is that a ray originally unpolarized loses energy by polarization by the factor $(1 + \cos^2 2\theta)/2$ as measured in the reflected ray. Most intensity data are taken from rotating crystals, and this introduces another factor, that must be corrected for. We have to deal with this factor, the Lorentz factor, because many reflections pass through the zone of reflection faster than others. This effect does not apply to our orientation observations because the crystal can be held stationary.

Appendix

The most methodical way of determining the orientation of a random plate of a known crystal is that of observing three noncoplanar X-ray reflections. The three noncoplanar unit vectors normal to the planes $(h_1k_1l_1)$, $(h_2k_2l_2)$, $(h_3k_3l_3)$ form a coordinate basis which we call the M (for Miller) basis. Our X-ray measurements give the relation between these atomic plane normals and the plate edges. From this we can calculate the plate thickness and length directions on the crystal axes. However, the actual unit cell edges are not, in general, the best basis for a coordinate system because elastic and electric field problems are most simple when stated on a basis of three mutually perpendicular unit axes X, Y, and Z.

In the most general unit cell, the triclinic, the basic vectors a, b, c have lengths a_0, b_0, c_0, respectively and make angles α, β, γ with each other, that is, c is at an angle α with b, a is at an angle β with c and b is at an angle γ with a. We call this the C (for unit cell) basis. We now set up a basis of axes X, Y, Z called the K (for Kristall) basis. Z lies along c, Y is perpendicular to c and a, and X is in the plane of a and c. X, Y, Z is an orthogonal unitary system.

We can convert vectors from the C basis to the K basis by the use of the relation

$$V_{(K)} = (K|C)V_{(c)}$$

where $(K|C)$ is the matrix

$$\begin{pmatrix} a_0 \sin \beta & b_0 u & 0 \\ 0 & b_0 v & 0 \\ a_0 \cos \beta & b_0 \cos \alpha & c_0 \end{pmatrix}$$

where

$$u = \frac{\cos \gamma - \cos \alpha \cos \beta}{\sin \beta} \quad \text{and} \quad v = \sqrt{\sin^2 \alpha - u^2}$$

The columns of $(K|C)$ are the vectors a, b, c, expressed on the K vectors X, Y, Z. We can reverse the relation to $V_{(c)} = (K|C)^{-1} V_{(k)}$ which converts any vector expressed on the K basis to its components on the C basis. Calculating the reciprocal of $(K|C)$ gives

$$(C|K) = \begin{pmatrix} \dfrac{1}{a_0 \sin \beta} & -\dfrac{u}{a_0 v \sin \beta} & 0 \\[2ex] 0 & \dfrac{1}{b_0 v} & 0 \\[2ex] -\dfrac{\cot \beta}{c_0} & \dfrac{-\cos \alpha + u \cot \beta}{c_0 v} & \dfrac{1}{c_0} \end{pmatrix}$$

The columns of $(C|K)$ are the axes X, Y, Z expressed on the C basis. The vector normal to the plane (hkl) is

$$(N_{hkl})_{(K)} = (\widetilde{C|K}) \begin{pmatrix} h \\ k \\ l \end{pmatrix}$$

where $(\widetilde{C|K})$ is the transpose of $(C|K)$. Hence we have

$$(N_{hkl})_{(K)} = \begin{pmatrix} \dfrac{h}{a_0 \sin \beta} - \dfrac{l \cot \beta}{c_0} \\[2ex] -\dfrac{hu}{a_0 v \sin \beta} + \dfrac{k}{b_0 v} - \dfrac{l \cos \alpha}{c_0 v} + \dfrac{ul \cot \beta}{c_0 v} \\[2ex] \dfrac{l}{c_0} \end{pmatrix}_{(K)}$$

This gives the three components on the K basis of the vector perpendicular to the (hkl) plane for a triclinic crystal. To form the matrix $(K|M)$ we take three such vectors and normalize them. The three column vectors form $(K|M)$.

For monoclinic crystals $\gamma = \alpha = 90°$ which makes $u = 0$ and $v = 1$. Hence the plane (hkl) has a normal

$$(N_{hkl})_{(K)} = \begin{pmatrix} \dfrac{h}{a_0 \sin \beta} - \dfrac{l \cot \beta}{c_0} \\[2ex] \dfrac{k}{b_0} \\[2ex] \dfrac{l}{c_0} \end{pmatrix}$$

Again, three of these, normalized, form $(K|M)$.

For orthorhombic crystals $\alpha = \beta = \gamma = 90°$ and we have

$$(N_{hkl})_{(K)} = \begin{pmatrix} h/a_0 \\ k/b_0 \\ l/c_0 \end{pmatrix}_{(K)}$$

In tetragonal crystals $b_0 = a_0$ giving

$$(N_{hkl})_{(K)} = \begin{pmatrix} h/a_0 \\ k/a_0 \\ l/c_0 \end{pmatrix}_{(K)}$$

For hexagonal crystals $\alpha = \beta = 90°$, $\gamma = 120°$ and $b_0 = a_0$, so that $u = -\frac{1}{2}$, $v = \sqrt{3}/2$ and

$$(N_{hkl})_{(K)} = \begin{pmatrix} \dfrac{h}{a_0} \\ \dfrac{h+2k}{\sqrt{3}a_0} \\ \dfrac{l}{c_0} \end{pmatrix}_{(K)}$$

For cubic crystals $\alpha = \beta = \gamma = 90°$ and $a_0 = b_0 = c_0$ and we have

$$(N_{hkl})_{(K)} = \frac{1}{a_0}\begin{pmatrix} h \\ k \\ l \end{pmatrix}_{(K)}$$

All these vectors need to be normalized before forming $(K|M)$ because when we observe the M vectors on the plate axes, we will assume them to be of unit length. Of course $V_{(K)} = (K|M)V_{(M)}$.

We now use our X-ray measurements to relate the Miller basis M to a unit orthogonal basis P. Here P_1 lies along the plate thickness direction, P_2 along the plate length direction, and P_3 along the plate width direction. We find a first reflection at a goniometer arm reading A_1, a counting tube reading $2\theta_1$ and at azimuth ϕ_1 as in Fig. A.1. For the accurate determination of ϕ it is well to use a collimator with pin holes instead of slits and to be able to place a horizontal slit over the vertical slit of the counting tube after a reflection has been found. This atomic reflecting plane obviously makes an angle $\delta_1 = A_1 - \theta_1$ with the major surface of the plate. Hence

$$(M_1)_{(P)} = \begin{pmatrix} \cos\delta_1 \\ \sin\delta_1 & \cos\phi_1 \\ \sin\delta_1 & \sin\phi_1 \end{pmatrix}_{(P)}$$

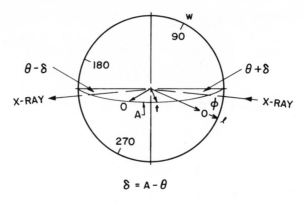

$$\delta = A - \theta$$

Figure A.1   Measuring A, θ, and ϕ to determine δ.

The value of θ generally establishes $(h_1 k_1 l_1)$. Bringing in the second and third X-ray measurements, we write $V_{(P)} = (P|M)V_{(M)}$ where

$$(P|M) = \begin{pmatrix} \cos \delta_1 & \cos \delta_2 & \cos \delta_3 \\ \sin \delta_1 \cos \phi_1 & \sin \delta_2 \cos \phi_2 & \sin \delta_3 \cos \phi_3 \\ \sin \delta_1 \sin \phi_1 & \sin \delta_2 \sin \phi_2 & \sin \delta_3 \sin \phi_3 \end{pmatrix}$$

We wish to evaluate $(K|P)$ because it gives the plate thickness and length directions on the K basis. From the three values of θ_i we choose a self-consistent set $(h_1 k_1 l_1)$, $(h_2 k_2 l_2)$, $(h_3 k_3 l_3)$ and calculate $(K|M)$. From $(P|M)$ we can calculate $(M|P) = (P|M)^{-1}$ then as

$$V_{(K)} = (K|M)V_{(M)} \quad \text{and} \quad V_{(M)} = (M|P)V_{(P)}$$

we have

$$V_{(K)} = (K|M)(M|P)V_{(P)}$$

That is,

$$(K|M)(M|P) = (K|P)$$

We illustrate with the example of an iodic acid plate. The orthorhombic cell has dimensions, $a_0 = 5.538$ Å, $b_0 = 5.888$ Å, $c_0 = 7.733$ Å. The first reflection was found at $A_1 = 31.27°$, $\theta_1 = 26.22°$, $\phi_1 = 28.1°$, hence $\delta_1 = A_1 - \theta_1 = 5.05°$. To get this accuracy we change ϕ by $180°$ and measure $\bar{A} = \theta - \delta$ then take $2\delta_1 = A_1 - \bar{A}_1$, $2\theta_1 = A + \bar{A}_1$. From the value of θ this must be the (114) plane. A second reflection is found at $A_2 = 33.32°$, $\theta_2 = 19.03°$, $\phi_2 = 227.8°$ giving $\delta_2 = 14.29°$ and is obviously an (013) plane. A third reflection is found at $A_3 = 39.07°$, $\phi_3 = 250.1°$, $\theta_3 = 24.78°$, $\delta_3 =$

14.29° and must be an (014) plane. From these readings we calculate

$$(P|M) = \begin{pmatrix} 0.996118 & 0.969059 & 0.969059 \\ 0.077649 & -0.165801 & -0.084016 \\ 0.041461 & -0.182853 & -0.232091 \end{pmatrix}$$

From (114), (013), and (014), we calculate $(K|M)$ to be

$$(K|M) = \begin{pmatrix} 0.314805 & 0 & 0 \\ 0.296092 & 0.401036 & 0.311952 \\ 0.901791 & 0.916062 & 0.950098 \end{pmatrix}$$

Now $(P|M)^{-1} = (M|P)$ and $(K|M)(M|P) = (K|P)$ the result we are after. Because of the two zeros in $(K|M)$ it is easier to calculate $(M|K) = (K|M)^{-1}$ then calculate $(P|M)(M|K) = (P|K)$. Since both P and K are orthogonal unitary systems, the transpose is equal to the reciprocal: $(P|K)^{-1} = (\widetilde{P|K}) = (K|P)$

$$(M|K) = \begin{pmatrix} 3.176565 & 0 & 0 \\ -0.000033 & 9.974154 & -3.274874 \\ -3.015029 & -9.616834 & 4.210076 \end{pmatrix}$$

$$(P|M)(M|K) = (P|K) = \begin{pmatrix} 0.242461 & 0.346264 & 0.906266 \\ 0.499973 & -0.845757 & 0.189264 \\ 0.831471 & 0.408177 & -0.378300 \end{pmatrix}$$

and

$$(K|P) = \begin{pmatrix} 0.242461 & 0.499973 & 0.831471 \\ 0.346264 & -0.845757 & 0.408177 \\ 0.906266 & 0.189264 & -0.378300 \end{pmatrix}$$

Hence t has direction cosines 0.242461, 0.346264, 0.906266; that is, it is 75.97° from X, 69.74° from Y, and 25.01° from Z. Similarly l is 60.00° from X, 147.75° from Y, and 79.09° from Z.

STEREOGRAPHIC PROJECTION

The unit vector $\begin{pmatrix} x \\ y \\ z \end{pmatrix}$ has a stereographic projection on the $x - y$ plane: $x_s = x/(1+z)$, $y_s = y/(1+z)$. Given x_s and y_s, we can compute $u = x_s^2 + y_s^2$. Now

$$z = \frac{1-u}{1+u}, \qquad x = \frac{2x_s}{1+u}, \qquad y = \frac{2y_s}{1+u}$$

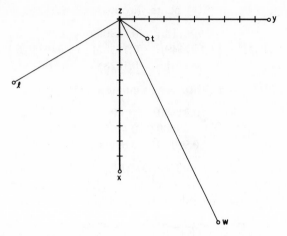

Figure A.2 Stereographic projection of t, l, and w on K basis, calculated from $(K|P)$.

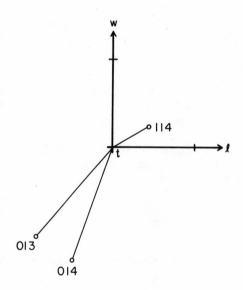

Figure A.3 Stereographic projection of (114), (013), and (014) plane normals on P basis, calculated from $(P|M)$.

The longitude of the point $[x, y, z]$ is

$$\tan^{-1} \frac{y}{x} = \tan^{-1} \frac{y_s}{x_s}$$

and the colatitude is $2 \tan^{-1} \sqrt{u}$. Using these relations we plot the above t, l, and w on the XY plane in Fig. A.2.

In Fig. A.3 we plot the normals to the (114) plane the (013) and the (014) on the plate axes. The projection is on the $l - w$ plane and was made from the $(P|M)$ matrix. The (114) direction has

$$y = \frac{0.077647}{1.996218}, \qquad z = \frac{0.041461}{1.99611} \cdots$$

This is drawn to a larger scale than was Fig. A.2.

Index